The Perfect Workbook – Math

Book 1: Addition and Subtraction

(Grades: K – 1)

First Edition

ISBN: 978-1-7353728-0-8

www.ThePerfectWorkbook.com

Table of Contents

Introduction

I am a mom of twins. When my children were ready to go to school, I wanted to find workbooks to supplement their math education at home. However, I couldn't find the "perfect workbook" for them. As a full-time attorney and true believer of practice-makes-perfect, I wanted a workbook from which I could assign a couple of pages per day that my children could do by themselves without too much hands-on instruction, and that would provide plenty of no-nonsense math questions (no number coloring!). Out of necessity, I created *The Perfect Workbook - Math.*

The Perfect Workbook - Math **provides:**

1. Easy Instructions: Like you, my days are busy and full, and I often find myself with very little time (and sometimes very little energy!) to be able to sit down to go over lessons and teach my children. I wanted a workbook that would provide simple and easy instructions for parents or caretakers to communicate to their children, as well as examples illustrating why math works the way it works, and then set daily assignments and goals for the kids. However, a lot of workbooks required heavy parental supervision and monitoring due to constantly changing or inconsistent instructions. ***The Perfect Workbook - Math*** **is designed to allow kids to independently think through the math problems and complete their assignments once a short key concept for each lesson is taught.**

2. Extensive Practice Questions: I grew up in a household where education was taken seriously. Based on my experience, I firmly believe in the value of practice. In order to excel at sports, play musical instruments, or apply any skill, there is no doubt that deliberate and rigorous practice is important. So why not math? Unfortunately, I could not find workbooks that provided sufficient practice questions for each lesson. ***The Perfect Workbook - Math*** **provides ample practice questions so that kids are equipped to master each lesson.**

3. Structured and Methodical Lessons: I purchased many math workbooks that contained a multitude of varying types of mathematics instruction. For example, one workbook would have addition and subtraction lessons, as well as lessons on measurement and shapes. I also purchased many math workbooks that only contained repetitive math practice questions for one type of arithmetic operation, such as multiplication. Both types of workbooks were not structured or methodical. For that reason, it was difficult to teach because I found myself jumping from one concept to another and trying to figure out what the next step should be. I wanted a workbook that would improve my children's math skills in a steady, logical, and incremental way, like climbing a staircase, and not like an obstacle course where things are introduced at random. ***The Perfect Workbook - Math*** **is organized in a step-by-step way, incrementally building on foundations.**

4. Fundamental Math Concepts: I strongly believe that the ultimate goal of learning math is understanding how to think analytically and critically to solve problems. The goal of learning math is not knowing how to master a six-digit division computation, but is rather appreciating how to think outside the box, separating a problem into parts, and seeing the world of quantity in multiple dimensions. This goal cannot be achieved if the fundamentals are not firmly established first. Skipping the fundamentals is akin to teaching a child the alphabet once and then asking the child to write a book. I wanted a standalone math workbook that could guide children through all of the fundamental math concepts at each grade level. **The Perfect Workbook - Math teaches the fundamental math concepts to equip the kids for the next level of math and so on.**

5. Tangible Learning Design: Nowadays, a lot of educational materials are screen-based and online. Although these electronic-based materials provide instant feedback and attention-grabbing graphics, I noticed that the materials are sometimes focused more on providing entertainment and focused less on allowing the kids to think through the questions; the kids become too eager to click buttons to find the answers, which comes easy, instead of actually working through and solving them. I wanted a paper-based workbook on which children could write on their own to solve questions and reflect tangible evidence of their efforts. I remember the first workbook that my kids finished. My husband and I recognized my kids' achievement and celebrated their hard work as a family. The kids were so proud of themselves and were ready to do it again. ***The Perfect Workbook - Math*** **provides a clean and thoughtful layout for each practice question so that kids can show their work.**

6. The Name "The Perfect Workbook": My friends and family advised me not to name the workbook, "The Perfect Workbook." They asked: "What if the workbook has a printing error?"; "What if someone finds the workbook less than perfect and criticizes your title?"; "Would you really want to put the standard that high?" Those questions made me pause and seriously reconsider the title. However, I couldn't let go of my commitment to create a workbook that is effective, beneficial, and perfect for children. This workbook is the product of my commitment. **I hope *The Perfect Workbook - Math* can help your child to learn to love math and build confidence in their abilities, as it did with mine.**

I am eternally grateful to my husband Ed. I would not have been able to create this workbook without his support and encouragement. I am also thankful to my friend Michelle, who offered great ideas through the process.

Sharon from The Perfect Workbook LLC
February 2021

Structure of the Perfect Workbook - Math

1. Each book of *The Perfect Workbook – Math* includes 10 units.

2. Each unit introduces a new concept. Each unit builds on the prior units, so children can learn math in a steady, logical, and incremental way.

3. Each unit begins with one page of lessons and examples. Please make sure to go over this lesson page before starting the following math practice questions.

4. Each unit includes 10 pages of math practice questions. The book is designed to provide 2 pages of practice questions (e.g., Day 1A and 1B) per weekday (Monday through Friday).

5. Each practice page starts with a question, "Why am I learning this?" Under that question, there is a math word question that can be solved by a concept related to that page. With this question, I hope that children can see how math can be fun and useful, and math can actually apply to everyday situations. The first numeric math practice question on each practice page gives an answer to the math word question.

6. The answer sheets are included at the back of the book.

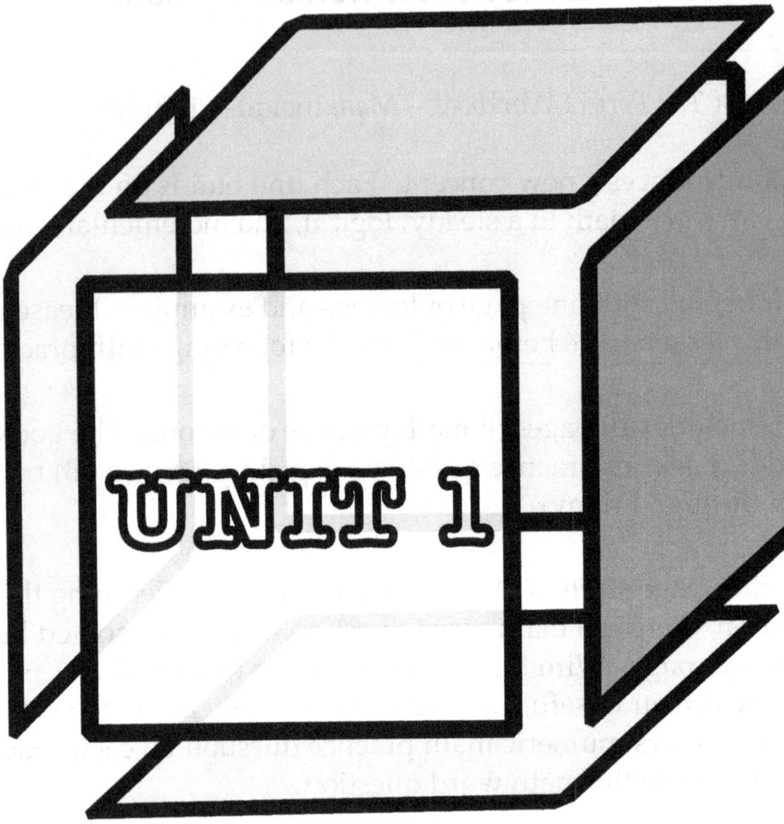

Addition up to 9

Unit 1: Addition up to 9

Lesson:

In this unit, we will learn about "**addition**." Addition means **putting together** different numbers to make one larger number. This larger number is called the "**sum**" or the "**total**."

We use the "+" sign to show addition and the "=" to show the sum.

If you have **3 donuts** and add (put together) **2 more donuts**, then how many donuts would you have **in total**?

You could think of it in pictures as:

OR

You could think of it in numbers as:

3 + 2 = 5

Examples:

Ben has **4 pennies** in his pocket and finds **2 more pennies**.
Ben **adds 4 pennies** and **2 pennies**, and will have **a total of 6 pennies**.

	4
+	**2**
	6

OR

4 + 2 = 6

Unit 1: Additon up to 9

Day 1A **Date:**

Why am I learning this?

Luke has 1 pet dragon and 1 pet unicorn. I can figure out how many pets Luke has.

1.	2.	3.	4.	5.
1 + 1	2 + 2	3 + 3	4 + 4	5 + 3

6.	7.	8.	9.	10.
3 + 1	4 + 3	2 + 5	6 + 2	5 + 1

11.	12.	13.	14.	15.
7 + 2	2 + 4	3 + 1	4 + 2	2 + 3

16.	17.	18.	19.	20.
8 + 0	0 + 7	4 + 5	5 + 2	3 + 5

Day 1B Date:

Why am I learning this?

A pirate ship carries two treasure chests: one chest has 5 gold bars and the other has 2 gold bars. I can figure out how many gold bars are on the pirate ship.

1. 5 + 2 = ☐	2. 2 + 2 = ☐
3. 3 + 4 = ☐	4. 7 + 1 = ☐
5. 7 + 2 = ☐	6. 4 + 3 = ☐
7. 0 + 9 = ☐	8. 6 + 2 = ☐
9. 3 + 3 = ☐	10. 5 + 3 = ☐
11. 2 + 4 = ☐	12. 2 + 6 = ☐
13. 6 + 3 = ☐	14. 1 + 5 = ☐
15. 4 + 1 = ☐	16. 4 + 4 = ☐
17. 1 + 6 = ☐	18. 5 + 4 = ☐
19. 8 + 0 = ☐	20. 4 + 3 = ☐

Unit 1: Additon up to 9

Day 2A **Date:**

Why am I learning this?

Scarlett lost 7 pairs of shoes yesterday and lost 2 more pairs today. I can figure out how many shoes Scarlett lost yesterday and today.

1. $7 + 2 =$

2. $8 + 1 =$

3. $2 + 1 =$

4. $6 + 3 =$

5. $4 + 2 =$

6. $5 + 1 =$

7. $4 + 4 =$

8. $7 + 1 =$

9. $1 + 7 =$

10. $8 + 0 =$

11. $5 + 4 =$

12. $0 + 3 =$

13. $3 + 4 =$

14. $2 + 6 =$

15. $6 + 2 =$

16. $5 + 3 =$

17. $4 + 5 =$

18. $3 + 5 =$

19. $2 + 3 =$

20. $1 + 4 =$

Day 2B Date:

Why am I learning this?

Colton baked 4 spinach cookies and baked 5 cauliflower cookies. I can figure out how many yummy cookies Colton baked in total.

1. 4 + 5 = ☐

2. 5 + 1 = ☐

3. 3 + 2 = ☐

4. 4 + 2 = ☐

5. 2 + 6 = ☐

6. 8 + 0 = ☐

7. 1 + 0 = ☐

8. 7 + 2 = ☐

9. 7 + 1 = ☐

10. 6 + 1 = ☐

11. 0 + 9 = ☐

12. 3 + 5 = ☐

13. 6 + 3 = ☐

14. 5 + 4 = ☐

15. 3 + 4 = ☐

16. 8 + 1 = ☐

17. 5 + 3 = ☐

18. 1 + 6 = ☐

19. 2 + 7 = ☐

20. 3 + 3 = ☐

Unit 1: Additon up to 9

Day 3A **Date:**

Why am I learning this?

Grace built 1 merry-go-round in the morning and built 7 more in the afternoon. I can figure out how many merry-go-rounds Grace built today.

1. $1 + 7 =$ ___

2. $5 + 3 =$ ___

3. $2 + 2 =$ ___

4. $4 + 5 =$ ___

5. $2 + 4 =$ ___

6. $2 + 0 =$ ___

7. $4 + 4 =$ ___

8. $3 + 1 =$ ___

9. $6 + 2 =$ ___

10. $3 + 2 =$ ___

11. $3 + 3 =$ ___

12. $4 + 2 =$ ___

13. $2 + 6 =$ ___

14. $4 + 3 =$ ___

15. $5 + 4 =$ ___

16. $5 + 2 =$ ___

17. $1 + 3 =$ ___

18. $5 + 1 =$ ___

19. $6 + 3 =$ ___

20. $3 + 5 =$ ___

Unit 1: Additon up to 9

Day 3B **Date:**

Why am I learning this?

Ian has 3 puppies, and his dad brought 2 more puppies for Ian. I can figure out how many puppies Ian has.

1. $3 + 2 =$ ☐
2. $2 + 5 =$ ☐
3. $1 + 3 =$ ☐
4. $4 + 5 =$ ☐
5. $5 + 3 =$ ☐
6. $6 + 1 =$ ☐
7. $7 + 2 =$ ☐
8. $8 + 1 =$ ☐
9. $0 + 9 =$ ☐
10. $1 + 5 =$ ☐
11. $2 + 4 =$ ☐
12. $3 + 3 =$ ☐
13. $2 + 6 =$ ☐
14. $5 + 4 =$ ☐
15. $6 + 2 =$ ☐
16. $7 + 1 =$ ☐
17. $2 + 3 =$ ☐
18. $3 + 5 =$ ☐
19. $4 + 4 =$ ☐
20. $5 + 0 =$ ☐

Unit 1: Additon up to 9

Day 4A **Date:**

Why am I learning this?

Aiden ate 1 pizza pie for breakfast and ate 6 more pizza pies for lunch. I can figure out how many pizza pies Aiden ate in total.

1. $1 + 6 =$ ___
2. $5 + 3 =$ ___
3. $6 + 0 =$ ___
4. $8 + 1 =$ ___
5. $5 + 1 =$ ___
6. $6 + 1 =$ ___
7. $3 + 5 =$ ___
8. $0 + 6 =$ ___
9. $1 + 8 =$ ___
10. $1 + 5 =$ ___
11. $4 + 2 =$ ___
12. $7 + 2 =$ ___
13. $3 + 4 =$ ___
14. $2 + 3 =$ ___
15. $2 + 6 =$ ___
16. $2 + 4 =$ ___
17. $2 + 7 =$ ___
18. $4 + 3 =$ ___
19. $3 + 2 =$ ___
20. $6 + 2 =$ ___

Unit 1: Additon up to 9

Day 4B **Date:**

Why am I learning this?

Zoe has 3 flying elephants, and her aunt gave 1 more flying elephant for Zoe. I can figure out how many flying elephants Zoe has.

1. 3 + 1 = ☐	2. 5 + 2 = ☐
3. 2 + 3 = ☐	4. 7 + 2 = ☐
5. 3 + 5 = ☐	6. 1 + 6 = ☐
7. 2 + 7 = ☐	8. 1 + 8 = ☐
9. 4 + 2 = ☐	10. 2 + 6 = ☐
11. 5 + 1 = ☐	12. 4 + 4 = ☐
13. 3 + 3 = ☐	14. 2 + 2 = ☐
15. 1 + 1 = ☐	16. 9 + 0 = ☐
17. 5 + 3 = ☐	18. 2 + 4 = ☐
19. 6 + 2 = ☐	20. 4 + 5 = ☐

Unit 1: Additon up to 9

Day 5A **Date:**

Why am I learning this?

Aurora lost 2 baby teeth last week, and lost 1 more baby tooth this week. I can figure out how many teeth Aurora has lost so far.

1. $2 + 1 =$ ___
2. $3 + 6 =$ ___
3. $2 + 2 =$ ___
4. $4 + 1 =$ ___
5. $2 + 3 =$ ___
6. $3 + 5 =$ ___
7. $2 + 4 =$ ___
8. $4 + 2 =$ ___
9. $2 + 5 =$ ___
10. $3 + 4 =$ ___
11. $2 + 6 =$ ___
12. $4 + 3 =$ ___
13. $2 + 7 =$ ___
14. $3 + 3 =$ ___
15. $4 + 4 =$ ___
16. $3 + 1 =$ ___
17. $5 + 2 =$ ___
18. $3 + 2 =$ ___
19. $5 + 1 =$ ___
20. $4 + 5 =$ ___

Day 5B Date:

Why am I learning this?

Sam invited 3 clowns for his birthday party, and the clowns came with 4 more clown friends to the party. I can figure out how many clowns came to Sam's birthday party.

1. 3 + 4 = ☐	2. 8 + 1 = ☐
3. 4 + 3 = ☐	4. 3 + 6 = ☐
5. 2 + 3 = ☐	6. 7 + 2 = ☐
7. 3 + 3 = ☐	8. 2 + 6 = ☐
9. 4 + 4 = ☐	10. 2 + 2 = ☐
11. 4 + 1 = ☐	12. 2 + 4 = ☐
13. 5 + 3 = ☐	14. 4 + 2 = ☐
15. 2 + 5 = ☐	16. 3 + 2 = ☐
17. 6 + 3 = ☐	18. 8 + 0 = ☐
19. 7 + 1 = ☐	20. 5 + 4 = ☐

UNIT 2

Subtraction up to 9

Unit 2: Subtraction up to 9

Lesson:

In this unit, we will learn about "**subtraction**." Subtraction means **taking away** a number from another number, also called the "difference," "what is remaining," or "what is left." Subtraction is the **opposite** of addition.

We use "–" to show subtraction and "=" to show the difference.

If you have **5 beach balls** and subtract (take away) **3 beach balls**, then how many beach balls would you have **left**?

You could think of it in pictures as:

OR

You could think of it in numbers as:

5 – 3 = 2

Examples:

Vivian had **9 gummy bears** in a jar and **ate 4** of them. Vivian **subtracts 4 gummy bears** from **9 gummy bears**, and will have **5 gummy bears remaining**.

	9
–	4
	5

OR

9 – 4 = 5

Unit 2: Subtraction up to 9

Day 1A **Date:**

Why am I learning this?

Elena had 4 coconuts and gave 3 coconuts to her best friend. I can figure out how many coconuts Elena has now.

1. $4 - 3 =$ ___
2. $8 - 2 =$ ___
3. $7 - 4 =$ ___
4. $6 - 6 =$ ___
5. $5 - 2 =$ ___
6. $7 - 3 =$ ___
7. $3 - 3 =$ ___
8. $2 - 1 =$ ___
9. $1 - 0 =$ ___
10. $9 - 4 =$ ___
11. $8 - 6 =$ ___
12. $7 - 5 =$ ___
13. $6 - 3 =$ ___
14. $5 - 1 =$ ___
15. $4 - 2 =$ ___
16. $3 - 2 =$ ___
17. $2 - 0 =$ ___
18. $1 - 1 =$ ___
19. $9 - 6 =$ ___
20. $9 - 8 =$ ___

Unit 2: Subtraction up to 9

Day 1B Date:

Why am I learning this?

Wyatt had 5 cherries in his hand and dropped 3 of the cherries. I can figure out how many cherries Wyatt has left in his hand.

1. $5 - 3 = \square$	2. $9 - 5 = \square$
3. $9 - 4 = \square$	4. $9 - 7 = \square$
5. $9 - 2 = \square$	6. $8 - 1 = \square$
7. $8 - 3 = \square$	8. $8 - 5 = \square$
9. $8 - 7 = \square$	10. $7 - 2 = \square$
11. $7 - 3 = \square$	12. $7 - 5 = \square$
13. $7 - 6 = \square$	14. $6 - 3 = \square$
15. $6 - 2 = \square$	16. $6 - 4 = \square$
17. $6 - 5 = \square$	18. $9 - 1 = \square$
19. $7 - 4 = \square$	20. $8 - 8 = \square$

Unit 2: Subtraction up to 9	
Day 2A	Date:

Why am I learning this?

There were 9 chickens in a chicken coop, and a fox came and ate 3 chickens. I can figure out how many chickens are left in the coop.

1. $9 - 3$
2. $4 - 2$
3. $2 - 2$
4. $3 - 1$
5. $5 - 4$
6. $6 - 3$
7. $4 - 0$
8. $9 - 7$
9. $5 - 1$
10. $7 - 4$
11. $3 - 2$
12. $9 - 4$
13. $6 - 2$
14. $5 - 0$
15. $8 - 4$
16. $7 - 3$
17. $6 - 5$
18. $8 - 6$
19. $9 - 3$
20. $5 - 3$

Unit 2: Subtraction up to 9

Day 2B **Date:**

Why am I learning this?

Andrew had $8 in his wallet, and he spent $3 to buy slime. I can figure out how much money Andrew has left in his wallet.

1. 8 − 3 = ☐	2. 5 − 5 = ☐
3. 7 − 5 = ☐	4. 3 − 1 = ☐
5. 9 − 6 = ☐	6. 4 − 1 = ☐
7. 7 − 6 = ☐	8. 5 − 2 = ☐
9. 7 − 3 = ☐	10. 6 − 1 = ☐
11. 8 − 2 = ☐	12. 9 − 2 = ☐
13. 8 − 5 = ☐	14. 8 − 6 = ☐
15. 6 − 3 = ☐	16. 5 − 1 = ☐
17. 2 − 1 = ☐	18. 7 − 2 = ☐
19. 8 − 4 = ☐	20. 9 − 5 = ☐

Day 3A Date:

Why am I learning this?

Savannah had 8 horses in her barn, and 2 of them ran away. I can figure out how many horses Savannah has left in her barn.

1. 8 - 2	2. 7 - 3	3. 5 - 4	4. 8 - 5	5. 9 - 1
6. 9 - 8	7. 4 - 1	8. 3 - 3	9. 8 - 7	10. 6 - 2
11. 4 - 3	12. 9 - 5	13. 8 - 3	14. 6 - 4	15. 5 - 3
16. 2 - 0	17. 3 - 2	18. 9 - 3	19. 8 - 4	20. 7 - 6

Unit 2: Subtraction up to 9

Day 3B **Date:**

Why am I learning this?

An apple tree had 7 apples on it, and 3 of the apples fell from the tree. I can figure out how many apples are still on the tree.

1. $7 - 3 = \square$	2. $2 - 2 = \square$
3. $8 - 7 = \square$	4. $1 - 0 = \square$
5. $9 - 6 = \square$	6. $7 - 5 = \square$
7. $9 - 9 = \square$	8. $9 - 4 = \square$
9. $3 - 0 = \square$	10. $6 - 5 = \square$
11. $7 - 1 = \square$	12. $9 - 2 = \square$
13. $6 - 3 = \square$	14. $7 - 4 = \square$
15. $5 - 2 = \square$	16. $9 - 8 = \square$
17. $9 - 7 = \square$	18. $9 - 1 = \square$
19. $4 - 1 = \square$	20. $6 - 2 = \square$

Unit 2: Subtraction up to 9

Day 4A **Date:**

Why am I learning this?

Jacob found 9 gold coins, and gave the 9 gold coins to his mom and dad. I can figure out how many gold coins Jacob has now.

1. $9 - 9 =$ ___

2. $8 - 6 =$ ___

3. $7 - 1 =$ ___

4. $7 - 6 =$ ___

5. $6 - 4 =$ ___

6. $5 - 2 =$ ___

7. $7 - 2 =$ ___

8. $8 - 4 =$ ___

9. $4 - 2 =$ ___

10. $3 - 3 =$ ___

11. $2 - 1 =$ ___

12. $8 - 1 =$ ___

13. $5 - 4 =$ ___

14. $9 - 2 =$ ___

15. $9 - 5 =$ ___

16. $7 - 3 =$ ___

17. $3 - 1 =$ ___

18. $2 - 0 =$ ___

19. $1 - 1 =$ ___

20. $8 - 3 =$ ___

Day 4B Date:

Why am I learning this?

There were 8 cookies in a cookie jar, and my brother ate 5 of them. I can figure out how many cookies are left in the cookie jar.

1. $8 - 5 = \square$
2. $8 - 2 = \square$
3. $7 - 4 = \square$
4. $7 - 6 = \square$
5. $6 - 3 = \square$
6. $6 - 6 = \square$
7. $9 - 6 = \square$
8. $6 - 4 = \square$
9. $5 - 4 = \square$
10. $5 - 2 = \square$
11. $4 - 3 = \square$
12. $3 - 2 = \square$
13. $1 - 1 = \square$
14. $9 - 1 = \square$
15. $5 - 3 = \square$
16. $8 - 6 = \square$
17. $9 - 7 = \square$
18. $7 - 5 = \square$
19. $9 - 3 = \square$
20. $6 - 2 = \square$

Unit 2: Subtraction up to 9

Day 5A **Date:**

Why am I learning this?

Olivia can invite 7 friends for a play date. Olivia already invited her best friends, Sarah and Michael. I can figure out how many more friends Olivia can invite.

1. $7 - 2 =$ ___
2. $6 - 1 =$ ___
3. $3 - 0 =$ ___
4. $5 - 1 =$ ___
5. $7 - 7 =$ ___
6. $9 - 8 =$ ___
7. $9 - 4 =$ ___
8. $7 - 5 =$ ___
9. $6 - 5 =$ ___
10. $4 - 1 =$ ___
11. $2 - 2 =$ ___
12. $9 - 6 =$ ___
13. $6 - 2 =$ ___
14. $5 - 1 =$ ___
15. $8 - 7 =$ ___
16. $3 - 1 =$ ___
17. $8 - 8 =$ ___
18. $1 - 0 =$ ___
19. $6 - 3 =$ ___
20. $8 - 5 =$ ___

Unit 2: Subtraction up to 9

Day 5B | Date:

Why am I learning this?

A bus had 8 passengers on the bus, and at a bus stop, 3 passengers got off the bus. I can figure out how many passengers are still left on the bus.

1. $8 - 3 = \square$
2. $8 - 1 = \square$
3. $4 - 3 = \square$
4. $9 - 4 = \square$
5. $5 - 5 = \square$
6. $5 - 3 = \square$
7. $6 - 1 = \square$
8. $4 - 0 = \square$
9. $9 - 7 = \square$
10. $8 - 5 = \square$
11. $9 - 2 = \square$
12. $9 - 6 = \square$
13. $8 - 6 = \square$
14. $8 - 8 = \square$
15. $6 - 4 = \square$
16. $7 - 4 = \square$
17. $4 - 4 = \square$
18. $4 - 2 = \square$
19. $6 - 2 = \square$
20. $9 - 0 = \square$

Understanding Addition and Subtraction

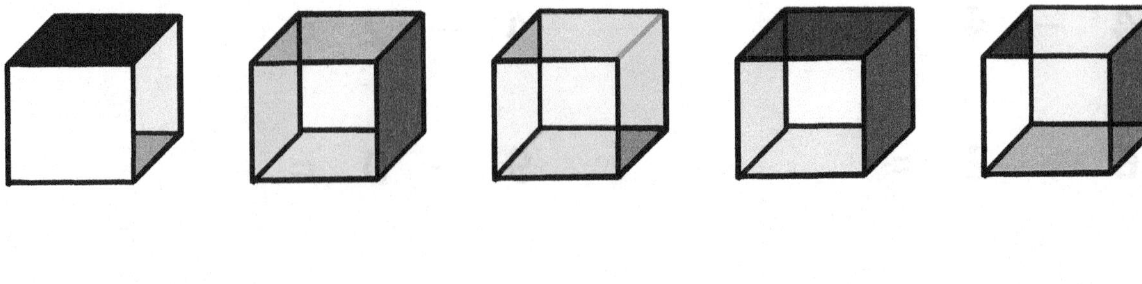

Unit 3: Understanding Addition and Subtraction

Lesson:

Addition and subtraction are opposites. That is because **addition is counting towards a total** and **subtraction is taking away from a total**. In this unit, we will see how addition and subtraction are related.

Examples:

Addition:

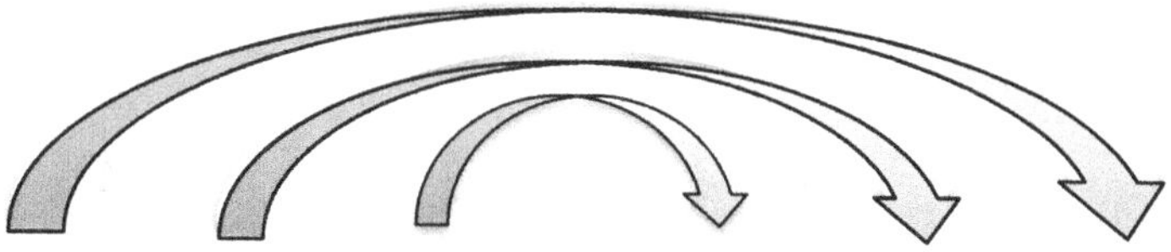

2 + 5 = 7   7 – 5 = 2

Subtraction:

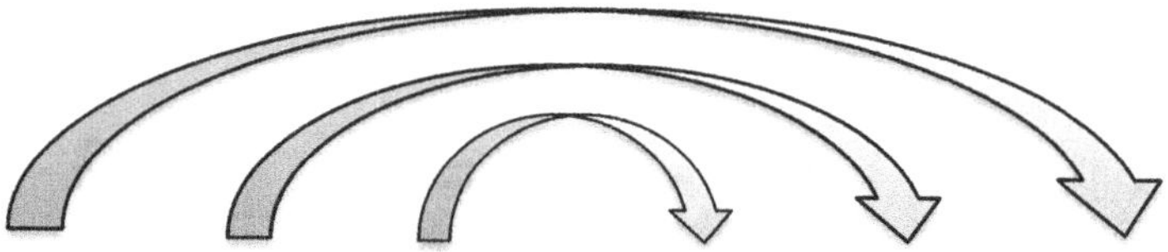

9 – 6 = 3   3 + 6 = 9

*Please note that understanding these addition and subtraction problems are PAIRED opposites, so it will be easy for students to sail through these. To help kids identify the patterns on their own, the following practice questions will add a third problem to mix things up. The goal is to keep them constantly thinking on their toes!

2 + 5 = 7	7 – 5 = 2	6 – 5 = 1
9 – 6 = 3	2 + 6 = 8	3 + 6 = 9

Unit 3: Understanding Addition and Subtraction

Day 1A Date:

Why am I learning this?

Eva had 2 balloons, and got 7 more ballons from her dad. I can figure out how many balloons Eva has. If 2 balloons float away, I can figure out how many balloons Eva has left.

1.	2 + 7 = ☐	9 - 2 = ☐	9 - 5 = ☐
2.	3 + 4 = ☐	6 - 3 = ☐	7 - 4 = ☐
3.	5 + 3 = ☐	5 - 5 = ☐	8 - 3 = ☐
4.	2 + 4 = ☐	6 - 2 = ☐	9 - 4 = ☐
5.	4 + 1 = ☐	7 - 4 = ☐	5 - 1 = ☐
6.	5 + 4 = ☐	6 - 5 = ☐	9 - 4 = ☐
7.	6 + 0 = ☐	6 - 6 = ☐	8 - 4 = ☐
8.	7 + 2 = ☐	8 - 7 = ☐	9 - 2 = ☐
9.	4 + 3 = ☐	7 - 4 = ☐	6 - 3 = ☐
10.	7 + 1 = ☐	8 - 7 = ☐	7 - 5 = ☐
11.	6 + 2 = ☐	7 - 3 = ☐	8 - 2 = ☐
12.	5 + 1 = ☐	6 - 5 = ☐	9 - 7 = ☐

Unit 3: Understanding Addition and Subtraction

Day 1B **Date:**

Why am I learning this?

Ben had 9 ponies and gave 5 ponies to his mom. I can figure out how many ponies Ben has left. If Ben's mom returns 5 ponies to Ben, I know how many ponies Ben has.

1.	9 - 5 =	4 + 5 =	2 + 5 =
2.	7 - 2 =	5 + 2 =	3 + 2 =
3.	6 - 5 =	5 + 3 =	1 + 5 =
4.	4 - 1 =	4 + 3 =	3 + 1 =
5.	5 - 2 =	3 + 2 =	6 + 2 =
6.	9 - 3 =	3 + 4 =	6 + 3 =
7.	6 - 5 =	6 + 2 =	0 + 5 =
8.	9 - 4 =	4 + 4 =	5 + 4 =
9.	7 - 1 =	6 + 1 =	4 + 1 =
10.	7 - 3 =	4 + 3 =	3 + 6 =
11.	8 - 5 =	2 + 3 =	3 + 5 =
12.	9 - 5 =	4 + 4 =	4 + 5 =

Unit 3: Understanding Addition and Subtraction

Day 2A **Date:**

Why am I learning this?

Jayden picked 3 apples and 5 pears. I can figure out how many fruits Jayden has. If Jayden eats all his pears, I can figure out how many fruits he has left.

1.	3 + 5 = ☐	8 - 5 = ☐	6 - 3 = ☐
2.	2 + 7 = ☐	6 - 5 = ☐	9 - 7 = ☐
3.	7 + 1 = ☐	3 - 2 = ☐	8 - 7 = ☐
4.	4 + 3 = ☐	7 - 3 = ☐	8 - 6 = ☐
5.	8 + 0 = ☐	8 - 8 = ☐	6 - 4 = ☐
6.	1 + 2 = ☐	3 - 2 = ☐	4 - 1 = ☐
7.	3 + 4 = ☐	7 - 4 = ☐	8 - 2 = ☐
8.	5 + 2 = ☐	4 - 2 = ☐	7 - 2 = ☐
9.	2 + 4 = ☐	5 - 0 = ☐	6 - 4 = ☐
10.	6 + 2 = ☐	8 - 6 = ☐	9 - 6 = ☐
11.	4 + 5 = ☐	2 - 2 = ☐	9 - 5 = ☐
12.	2 + 3 = ☐	6 - 2 = ☐	5 - 3 = ☐

Unit 3: Understanding Addition and Subtraction

Day 2B **Date:**

Why am I learning this?

Julia had 8 pet platypuses but 3 of them ran away. I can figure out how many platypuses Julia has left. If the 3 platypuses come back, I know how many platypuses Julia has.

1.	8 - 3 =	5 + 3 =	4 + 4 =
2.	5 - 3 =	2 + 6 =	2 + 3 =
3.	4 - 1 =	3 + 1 =	4 + 5 =
4.	9 - 7 =	3 + 3 =	2 + 7 =
5.	6 - 2 =	2 + 0 =	4 + 2 =
6.	7 - 4 =	3 + 4 =	3 + 6 =
7.	3 - 2 =	5 + 3 =	1 + 2 =
8.	9 - 4 =	5 + 4 =	2 + 6 =
9.	7 - 5 =	2 + 2 =	2 + 5 =
10.	5 - 1 =	4 + 1 =	4 + 3 =
11.	9 - 3 =	2 + 3 =	6 + 3 =
12.	6 - 1 =	1 + 4 =	5 + 1 =

Unit 3: Understanding Addition and Subtraction

Day 3A **Date:**

Why am I learning this?

There was 1 red bird sitting on a branch, and 3 blue birds came to sit with the red bird. If the red bird flies away, I know how many birds will be left on the branch.

1.	1 + 3 = ☐	4 - 1 = ☐	5 - 2 = ☐
2.	2 + 6 = ☐	8 - 2 = ☐	7 - 2 = ☐
3.	3 + 2 = ☐	6 - 1 = ☐	5 - 2 = ☐
4.	5 + 2 = ☐	9 - 2 = ☐	7 - 2 = ☐
5.	3 + 6 = ☐	8 - 4 = ☐	9 - 6 = ☐
6.	4 + 5 = ☐	9 - 5 = ☐	4 - 4 = ☐
7.	8 + 0 = ☐	8 - 8 = ☐	6 - 0 = ☐
8.	4 + 2 = ☐	5 - 4 = ☐	6 - 2 = ☐
9.	2 + 3 = ☐	5 - 3 = ☐	4 - 3 = ☐
10.	1 + 5 = ☐	8 - 6 = ☐	6 - 5 = ☐
11.	2 + 1 = ☐	3 - 1 = ☐	7 - 5 = ☐
12.	1 + 4 = ☐	7 - 4 = ☐	5 - 4 = ☐

Day 3B Date:

Why am I learning this?

Max had 7 invisible cloaks and gave 3 to his friends. I can figure out how many invisible cloaks Max has left. If Max's friends return 3 invisible cloaks, I know how many invisible cloaks Max has in total.

1.	7 - 3 = ☐	4 + 3 = ☐	1 + 5 = ☐
2.	4 - 1 = ☐	4 + 5 = ☐	3 + 1 = ☐
3.	8 - 5 = ☐	3 + 5 = ☐	2 + 2 = ☐
4.	6 - 2 = ☐	3 + 2 = ☐	4 + 2 = ☐
5.	3 - 2 = ☐	2 + 3 = ☐	1 + 2 = ☐
6.	5 - 2 = ☐	4 + 4 = ☐	3 + 2 = ☐
7.	9 - 4 = ☐	5 + 2 = ☐	4 + 5 = ☐
8.	9 - 6 = ☐	3 + 6 = ☐	6 + 2 = ☐
9.	7 - 5 = ☐	2 + 5 = ☐	4 + 4 = ☐
10.	5 - 4 = ☐	1 + 4 = ☐	2 + 7 = ☐
11.	8 - 3 = ☐	6 + 3 = ☐	3 + 3 = ☐
12.	6 - 4 = ☐	2 + 4 = ☐	4 + 0 = ☐

Unit 3: Understanding Addition and Subtraction	
Day 4A	Date:

Why am I learning this?

Tooth fairy collected 4 teeth from one child and 3 teeth from another child. I can figure out how many teeth that the Tooth fairy has. If Tooth fairy throws out 3 teeth because they are rotten, I can figure out how many teech Tooth fairy has.

1.	4 + 3 = ☐	7 - 3 = ☐	4 - 3 = ☐
2.	1 + 8 = ☐	5 - 3 = ☐	9 - 8 = ☐
3.	3 + 2 = ☐	8 - 5 = ☐	5 - 2 = ☐
4.	0 + 4 = ☐	4 - 4 = ☐	9 - 4 = ☐
5.	5 + 3 = ☐	9 - 2 = ☐	8 - 3 = ☐
6.	4 + 5 = ☐	6 - 5 = ☐	9 - 5 = ☐
7.	2 + 5 = ☐	7 - 5 = ☐	7 - 2 = ☐
8.	8 + 0 = ☐	8 - 8 = ☐	6 - 2 = ☐
9.	6 + 2 = ☐	9 - 5 = ☐	8 - 2 = ☐
10.	2 + 7 = ☐	7 - 3 = ☐	9 - 7 = ☐
11.	2 + 4 = ☐	9 - 3 = ☐	6 - 4 = ☐
12.	7 + 1 = ☐	8 - 7 = ☐	8 - 1 = ☐

Day 4B Date:

Why am I learning this?

Wizard had 9 spell books but spilled evil potion over 2 books and ruined them. I can figure out how many spell books are left for Wizard. If Elf gave 2 new spell books to Wizard, I can figure out how many spell books Wizard has.

1.	9 - 2 = ☐	7 + 2 = ☐	4 + 4 = ☐
2.	5 - 4 = ☐	3 + 6 = ☐	1 + 4 = ☐
3.	8 - 3 = ☐	5 + 3 = ☐	2 + 5 = ☐
4.	7 - 4 = ☐	3 + 4 = ☐	2 + 4 = ☐
5.	4 - 3 = ☐	1 + 3 = ☐	4 + 4 = ☐
6.	6 - 5 = ☐	1 + 5 = ☐	3 + 2 = ☐
7.	8 - 2 = ☐	2 + 6 = ☐	6 + 3 = ☐
8.	9 - 5 = ☐	2 + 7 = ☐	4 + 5 = ☐
9.	7 - 5 = ☐	2 + 5 = ☐	3 + 1 = ☐
10.	5 - 3 = ☐	5 + 0 = ☐	2 + 3 = ☐
11.	3 - 2 = ☐	1 + 2 = ☐	3 + 3 = ☐
12.	8 - 7 = ☐	2 + 2 = ☐	1 + 7 = ☐

Day 5A Date:

Why am I learning this?

Ava picked 5 carrots from her garden on Monday, and 3 carrots on Tuesday. I can figure out how many carrots Ava picked. If Ava eats 3 carrots on Thursday, I can figure out how many carrots Ava has left.

1.	5 + 3 = ☐	9 - 2 = ☐	8 - 3 = ☐
2.	4 + 5 = ☐	4 - 4 = ☐	9 - 5 = ☐
3.	3 + 6 = ☐	5 - 4 = ☐	9 - 6 = ☐
4.	2 + 3 = ☐	5 - 3 = ☐	8 - 7 = ☐
5.	1 + 5 = ☐	4 - 3 = ☐	6 - 5 = ☐
6.	6 + 2 = ☐	7 - 5 = ☐	8 - 2 = ☐
7.	7 + 1 = ☐	8 - 1 = ☐	7 - 2 = ☐
8.	8 + 1 = ☐	3 - 1 = ☐	9 - 1 = ☐
9.	9 + 0 = ☐	9 - 9 = ☐	7 - 6 = ☐
10.	5 + 4 = ☐	8 - 2 = ☐	9 - 4 = ☐
11.	4 + 2 = ☐	5 - 4 = ☐	6 - 2 = ☐
12.	3 + 4 = ☐	7 - 3 = ☐	7 - 4 = ☐

Day 5B Date:

Why am I learning this?

Kai played with 5 friends but 2 of them had to leave. I can figure out how many friends Kai has left to play with. If 2 of his friends come back, I can figure out how many friends Kai has in total to play with.

1.	5 - 2 = ☐	3 + 2 = ☐	2 + 3 = ☐
2.	9 - 7 = ☐	3 + 4 = ☐	2 + 7 = ☐
3.	2 - 0 = ☐	2 + 0 = ☐	3 + 6 = ☐
4.	6 - 4 = ☐	2 + 4 = ☐	1 + 1 = ☐
5.	8 - 5 = ☐	1 + 4 = ☐	3 + 5 = ☐
6.	3 - 3 = ☐	0 + 3 = ☐	5 + 2 = ☐
7.	7 - 6 = ☐	2 + 2 = ☐	1 + 6 = ☐
8.	9 - 4 = ☐	5 + 4 = ☐	6 + 1 = ☐
9.	8 - 2 = ☐	6 + 2 = ☐	2 + 6 = ☐
10.	7 - 4 = ☐	3 + 3 = ☐	3 + 4 = ☐
11.	5 - 4 = ☐	5 + 3 = ☐	1 + 4 = ☐
12.	9 - 3 = ☐	4 + 4 = ☐	6 + 3 = ☐

UNIT 4

Continuous Addition and Subtraction

Unit 4: Continuous Addition and Subtraction

Lesson:

In this unit, we will learn how to add and subtract three numbers. You already know how to do this! You can add and subtract numbers in the order they appear.

Examples:

Addition:

2 + 5 + 1 = ?

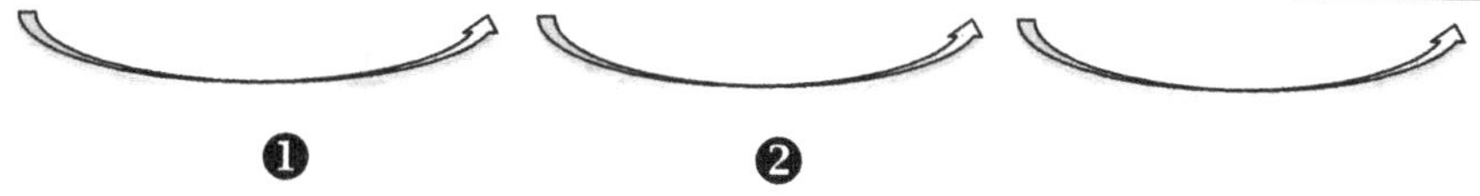

❶ 2 + 5 = 7

❷ 7 + 1 = 8

The answer is 8!

Subtraction:

9 − 4 − 2 = ?

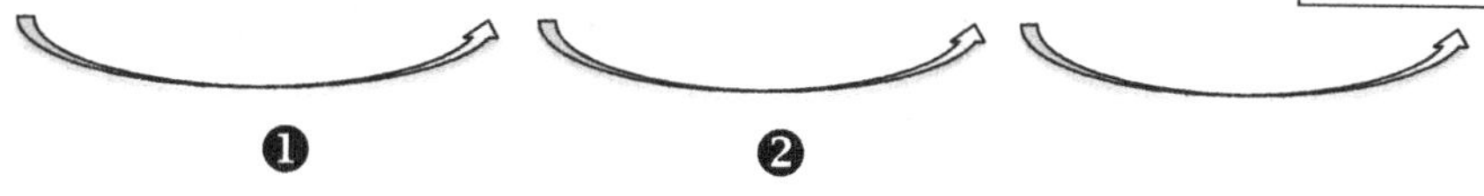

❶ 9 − 4 = 5

❷ 5 − 2 = 3

The answer is 3!

Unit 4: Continuous Addition and Subtraction

Day 1A **Date:**

Why am I learning this?

David had 1 seashell in his pocket and found 1 more seashell. David's mom then gave 4 more seashells to David. I can figure out how many total seashells David has.

1. 1 + 1 + 4 = ☐	2. 2 + 2 + 6 = ☐
3. 3 + 3 + 3 = ☐	4. 4 + 4 + 2 = ☐
5. 5 + 3 + 0 = ☐	6. 3 + 1 + 6 = ☐
7. 8 + 1 + 0 = ☐	8. 1 + 2 + 5 = ☐
9. 3 + 1 + 4 = ☐	10. 2 + 5 + 2 = ☐
11. 4 + 1 + 5 = ☐	12. 7 + 1 + 2 = ☐
13. 1 + 8 + 1 = ☐	14. 0 + 7 + 2 = ☐
15. 3 + 2 + 4 = ☐	16. 1 + 3 + 1 = ☐
17. 4 + 2 + 2 = ☐	18. 4 + 3 + 2 = ☐
19. 1 + 8 + 0 = ☐	20. 2 + 0 + 3 = ☐

Unit 4: Continuous Addition and Subtraction

Day 1B **Date:**

Why am I learning this?

Sofia has $9 and wants to buy a notebook that costs $4 and a pen that costs $1. Sofia buys a notebook and pen at the store with her money. I can figure out how much money Sofia has left.

1. 9 - 4 - 1 = ☐
2. 8 - 2 - 1 = ☐
3. 7 - 4 - 3 = ☐
4. 6 - 1 - 3 = ☐
5. 5 - 2 - 1 = ☐
6. 4 - 3 - 1 = ☐
7. 9 - 3 - 4 = ☐
8. 9 - 6 - 1 = ☐
9. 10 - 4 - 4 = ☐
10. 9 - 4 - 3 = ☐
11. 8 - 6 - 0 = ☐
12. 10 - 7 - 2 = ☐
13. 6 - 3 - 2 = ☐
14. 8 - 1 - 4 = ☐
15. 5 - 2 - 2 = ☐
16. 7 - 2 - 3 = ☐
17. 10 - 5 - 2 = ☐
18. 5 - 3 - 2 = ☐
19. 9 - 2 - 3 = ☐
20. 7 - 2 - 4 = ☐

Unit 4: Continuous Addition and Subtraction

Day 2A **Date:**

Why am I learning this?

Gavin bought 3 boats in September, bought 1 more boat in October, and bought 5 more boats in November. I can figure out how many boats Gavin bought in those three months.

1. $3 + 1 + 5 = \square$
2. $2 + 2 + 1 = \square$
3. $2 + 4 + 3 = \square$
4. $5 + 4 + 1 = \square$
5. $1 + 3 + 0 = \square$
6. $1 + 7 + 1 = \square$
7. $0 + 3 + 0 = \square$
8. $4 + 2 + 4 = \square$
9. $2 + 1 + 2 = \square$
10. $0 + 8 + 1 = \square$
11. $4 + 3 + 2 = \square$
12. $1 + 2 + 4 = \square$
13. $3 + 3 + 1 = \square$
14. $1 + 5 + 2 = \square$
15. $3 + 2 + 3 = \square$
16. $1 + 7 + 0 = \square$
17. $1 + 3 + 3 = \square$
18. $5 + 1 + 2 = \square$
19. $6 + 1 + 2 = \square$
20. $1 + 4 + 5 = \square$

Unit 4: Continuous Addition and Subtraction

Day 2B **Date:**

Why am I learning this?

Cora had 10 hats, but lost 2 yesterday and then lost 4 more of them today. I can figure out how many hats Cora has left.

1. 10 - 2 - 4 = ☐	2. 7 - 4 - 1 = ☐
3. 8 - 2 - 3 = ☐	4. 5 - 1 - 1 = ☐
5. 6 - 5 - 1 = ☐	6. 10 - 3 - 4 = ☐
7. 9 - 2 - 6 = ☐	8. 9 - 1 - 3 = ☐
9. 10 - 2 - 5 = ☐	10. 3 - 0 - 1 = ☐
11. 7 - 1 - 4 = ☐	12. 9 - 2 - 2 = ☐
13. 8 - 1 - 2 = ☐	14. 8 - 6 - 1 = ☐
15. 7 - 5 - 1 = ☐	16. 5 - 1 - 3 = ☐
17. 10 - 2 - 2 = ☐	18. 8 - 4 - 2 = ☐
19. 6 - 0 - 3 = ☐	20. 8 - 5 - 2 = ☐

Unit 4: Continuous Addition and Subtraction

Day 3A | **Date:**

Why am I learning this?

Luca has 4 dinosaurs, and the dinosaurs laid 2 eggs last week and laid 3 eggs yesterday. All of the eggs hatched today. I can figure out how many dinosaurs Luca has in total.

1. $4 + 2 + 3 =$ ☐
2. $8 + 0 + 1 =$ ☐
3. $2 + 2 + 2 =$ ☐
4. $3 + 3 + 3 =$ ☐
5. $4 + 1 + 3 =$ ☐
6. $2 + 4 + 1 =$ ☐
7. $6 + 3 + 1 =$ ☐
8. $7 + 1 + 1 =$ ☐
9. $2 + 3 + 2 =$ ☐
10. $6 + 1 + 1 =$ ☐
11. $1 + 1 + 2 =$ ☐
12. $1 + 5 + 2 =$ ☐
13. $3 + 4 + 1 =$ ☐
14. $3 + 3 + 2 =$ ☐
15. $4 + 0 + 3 =$ ☐
16. $7 + 1 + 2 =$ ☐
17. $6 + 0 + 3 =$ ☐
18. $5 + 2 + 2 =$ ☐
19. $3 + 2 + 2 =$ ☐
20. $1 + 4 + 2 =$ ☐

Unit 4: Continuous Addition and Subtraction

Day 3B **Date:**

Why am I learning this?

Nolen had 9 horses. Nolen gave 2 of the horses to his dad and 3 of the horses to his mom. I can figure out how many horses Nolen has left for himself.

1. 9 - 2 - 3 = ☐	2. 6 - 3 - 3 = ☐
3. 8 - 7 - 0 = ☐	4. 5 - 2 - 2 = ☐
5. 7 - 1 - 2 = ☐	6. 10 - 4 - 4 = ☐
7. 8 - 3 - 4 = ☐	8. 9 - 4 - 3 = ☐
9. 10 - 3 - 5 = ☐	10. 8 - 5 - 1 = ☐
11. 9 - 6 - 2 = ☐	12. 9 - 1 - 3 = ☐
13. 7 - 3 - 2 = ☐	14. 9 - 3 - 3 = ☐
15. 5 - 1 - 1 = ☐	16. 6 - 1 - 4 = ☐
17. 10 - 4 - 5 = ☐	18. 8 - 0 - 2 = ☐
19. 6 - 2 - 3 = ☐	20. 7 - 2 - 0 = ☐

Unit 4: Continuous Addition and Subtraction

Day 4A **Date:**

Why am I learning this?

Claire had 1 gold coin, found another gold coin at the playground, and found another gold coin in the basement. I can figure out how many gold coins Claire has in total.

1. 1 + 1 + 1 = ☐	2. 2 + 2 + 3 = ☐
3. 4 + 3 + 2 = ☐	4. 1 + 1 + 7 = ☐
5. 4 + 2 + 3 = ☐	6. 3 + 1 + 1 = ☐
7. 6 + 1 + 1 = ☐	8. 2 + 6 + 1 = ☐
9. 2 + 7 + 1 = ☐	10. 6 + 0 + 2 = ☐
11. 5 + 1 + 3 = ☐	12. 1 + 2 + 3 = ☐
13. 4 + 0 + 1 = ☐	14. 6 + 3 + 1 = ☐
15. 2 + 0 + 4 = ☐	16. 5 + 3 + 2 = ☐
17. 1 + 0 + 3 = ☐	18. 2 + 3 + 2 = ☐
19. 3 + 3 + 1 = ☐	20. 1 + 8 + 1 = ☐

Unit 4: Continuous Addition and Subtraction

Day 4B **Date:**

Why am I learning this?

Robert had 6 hoverboards. He gave 1 of the hoverboards to his teacher, and gave 3 to his best friend. I can figure out how many hoverboards Robert has left.

1. 6 - 1 - 3 = ☐	2. 10 - 4 - 3 = ☐
3. 9 - 2 - 5 = ☐	4. 9 - 4 - 2 = ☐
5. 5 - 1 - 3 = ☐	6. 8 - 0 - 4 = ☐
7. 10 - 2 - 4 = ☐	8. 6 - 2 - 3 = ☐
9. 5 - 1 - 4 = ☐	10. 6 - 2 - 1 = ☐
11. 5 - 2 - 1 = ☐	12. 2 - 1 - 1 = ☐
13. 7 - 4 - 2 = ☐	14. 10 - 4 - 5 = ☐
15. 5 - 2 - 1 = ☐	16. 6 - 3 - 2 = ☐
17. 10 - 5 - 2 = ☐	18. 5 - 3 - 0 = ☐
19. 7 - 3 - 3 = ☐	20. 8 - 5 - 1 = ☐

Unit 4: Continuous Addition and Subtraction

Day 5A   Date:

Why am I learning this?

Chloe had 2 robots already. She built 1 more robot last week, and then built 4 more robots this week. I can figure out how many robots Chloe now has in total.

1. 2 + 1 + 4 = ☐
2. 1 + 2 + 3 = ☐
3. 5 + 3 + 2 = ☐
4. 3 + 0 + 7 = ☐
5. 5 + 1 + 1 = ☐
6. 4 + 1 + 3 = ☐
7. 5 + 1 + 2 = ☐
8. 1 + 3 + 2 = ☐
9. 3 + 3 + 1 = ☐
10. 1 + 6 + 0 = ☐
11. 2 + 5 + 3 = ☐
12. 4 + 1 + 2 = ☐
13. 2 + 0 + 4 = ☐
14. 5 + 2 + 2 = ☐
15. 3 + 0 + 4 = ☐
16. 4 + 1 + 5 = ☐
17. 2 + 2 + 4 = ☐
18. 3 + 3 + 3 = ☐
19. 5 + 3 + 1 = ☐
20. 1 + 4 + 1 = ☐

Unit 4: Continuous Addition and Subtraction

Day 5B Date:

Why am I learning this?

Emma had 6 rockets and launched 1 rocket yesterday, and then launched 3 more rocket today. I can figure out how many rockets Emma will have left after today.

1. $6 - 1 - 3 =$ ☐
2. $10 - 4 - 3 =$ ☐
3. $9 - 2 - 5 =$ ☐
4. $9 - 4 - 2 =$ ☐
5. $5 - 1 - 3 =$ ☐
6. $8 - 0 - 4 =$ ☐
7. $10 - 2 - 4 =$ ☐
8. $6 - 2 - 3 =$ ☐
9. $5 - 1 - 4 =$ ☐
10. $6 - 2 - 1 =$ ☐
11. $5 - 2 - 2 =$ ☐
12. $2 - 1 - 1 =$ ☐
13. $7 - 4 - 2 =$ ☐
14. $9 - 4 - 5 =$ ☐
15. $5 - 2 - 1 =$ ☐
16. $6 - 3 - 2 =$ ☐
17. $10 - 6 - 2 =$ ☐
18. $5 - 1 - 0 =$ ☐
19. $7 - 3 - 3 =$ ☐
20. $8 - 2 - 5 =$ ☐

UNIT 5

Relationship between Addition and Subtraction

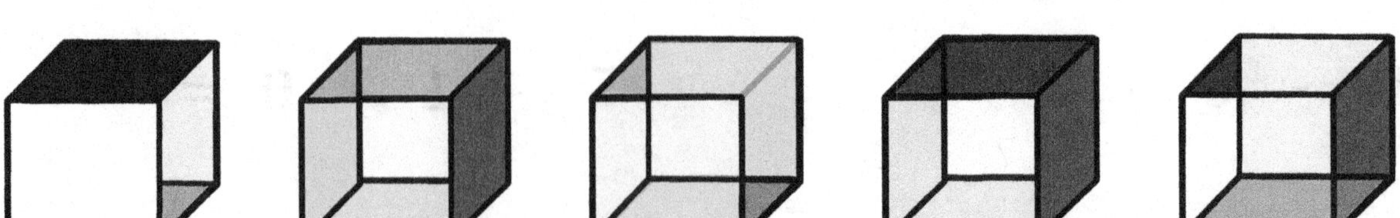

Unit 5: Relationship between Addition and Subtraction

Lesson:

In this unit, we will learn how addition and subtraction are related. Once you understand the relationship, you will be able to easily use addition and subtraction to solve a problem.

Because addition and subtraction are opposites, you can find a missing number in addition by using subtraction. Alternatively, you can find a missing number in subtraction by using addition. Can you see how addition and subtraction are related?

Examples:

①

$10 = 2 + \boxed{?}$

$\boxed{10 - 2 = 8}$ The answer is 8!

②

$8 = 4 + 3 + \boxed{?}$

$\boxed{8 - 4 - 3 = 1}$ The answer is 1!

③

$3 = 9 - \boxed{?}$

$\boxed{9 - 3 = 6}$ The answer is 6!

④

$2 = 6 - 3 - \boxed{?}$

$\boxed{6 - 3 - 2 = 1}$ The answer is 1!

Day 1A Date:

Why am I learning this?

Leo wants to collect 10 smooth pebbles. If Leo collected 1 pebble yesterday, I can figure out how many more pebbles that he wants to collect.

1. 10 = 1 + ☐

2. 10 = 2 + 5 + ☐

3. 10 = 4 + 3 + ☐

4. 10 = 3 + ☐

5. 10 = 5 + ☐

6. 10 = 4 + 4 + ☐

7. 10 = 8 + 1 + ☐

8. 10 = 7 + ☐

9. 10 = 9 + ☐

10. 10 = 1 + 7 + ☐

11. 10 = 3 + 3 + ☐

12. 10 = 2 + ☐

13. 10 = 4 + ☐

14. 10 = 5 + 2 + ☐

15. 10 = 7 + 3 + ☐

16. 10 = 6 + ☐

17. 10 = 8 + ☐

18. 10 = 1 + 8 + ☐

19. 10 = 0 + 5 + ☐

20. 10 = 0 + ☐

Unit 5: Relationship between Addition and Subtraction

Day 1B **Date:**

Why am I learning this?

Sadie wants to build 9 birdhouses. If Sadie made 1 birdhouse yesterday, I can figure out how many more birdhouses that Sadie wants to make.

1. $9 = 1 + \square$
2. $9 = 2 + 2 + \square$
3. $9 = 4 + 5 + \square$
4. $9 = 3 + \square$
5. $9 = 5 + \square$
6. $9 = 6 + 2 + \square$
7. $9 = 5 + 1 + \square$
8. $9 = 7 + \square$
9. $9 = 2 + \square$
10. $9 = 3 + 5 + \square$
11. $9 = 5 + 2 + \square$
12. $9 = 4 + \square$
13. $9 = 6 + \square$
14. $9 = 8 + 1 + \square$
15. $9 = 4 + 2 + \square$
16. $9 = 8 + \square$
17. $9 = 0 + \square$
18. $9 = 3 + 2 + \square$
19. $9 = 3 + 4 + \square$
20. $9 = 2 + 2 + \square$

Day 2A Date:

Why am I learning this?

Logan's mom asked Logan to pick 8 green peppers from their garden. Logan picked 1 green pepper so far. I can figure out how many more green peppers that Logan needs to pick.

1. $8 = 1 + \square$

2. $8 = 4 + 3 + \square$

3. $8 = 1 + 5 + \square$

4. $8 = 4 + \square$

5. $8 = 3 + \square$

6. $8 = 6 + 1 + \square$

7. $8 = 3 + 3 + \square$

8. $8 = 6 + \square$

9. $8 = 5 + \square$

10. $8 = 0 + 2 + \square$

11. $8 = 5 + 1 + \square$

12. $8 = 0 + \square$

13. $8 = 7 + \square$

14. $8 = 2 + 1 + \square$

15. $8 = 7 + 0 + \square$

16. $8 = 1 + \square$

17. $8 = 2 + \square$

18. $8 = 1 + 4 + \square$

19. $8 = 2 + 2 + \square$

20. $8 = 8 + \square$

Unit 5: Relationship between Addition and Subtraction

Day 2B **Date:**

Why am I learning this?

Lily wants to bake 7 coconut cakes. If Lily baked 1 cake yesterday, I can figure out how many more coconut cakes that Lily wants to bake.

1. 7 = 1 + □
2. 7 = 0 + 4 + □
3. 7 = 1 + 2 + □
4. 7 = 0 + □
5. 7 = 3 + □
6. 7 = 5 + 2 + □
7. 7 = 3 + 1 + □
8. 7 = 7 + □
9. 7 = 5 + □
10. 7 = 1 + 1 + □
11. 7 = 5 + 1 + □
12. 7 = 4 + □
13. 7 = 2 + □
14. 7 = 2 + 2 + □
15. 7 = 2 + 4 + □
16. 7 = 3 + 3 + □
17. 7 = 6 + □
18. 7 = 3 + 0 + □
19. 7 = 6 + 1 + □
20. 7 = 4 + 1 + □

Unit 5: Relationship between Addition and Subtraction

Day 3A **Date:**

Why am I learning this?

James wants to build 6 submarines. If James built 1 submarine yesterday, I can figure out how many more submarines James wants to build.

1. $6 = 1 + \square$
2. $6 = 0 + 2 + \square$
3. $6 = 1 + 4 + \square$
4. $6 = 0 + \square$
5. $6 = 3 + \square$
6. $6 = 1 + 2 + \square$
7. $6 = 3 + 1 + \square$
8. $6 = 6 + \square$
9. $6 = 5 + \square$
10. $6 = 1 + 3 + \square$
11. $6 = 5 + 1 + \square$
12. $6 = 3 + 3 + \square$
13. $6 = 2 + \square$
14. $6 = 0 + 4 + \square$
15. $6 = 2 + 1 + \square$
16. $6 = 1 + 1 + \square$
17. $6 = 4 + \square$
18. $6 = 3 + 2 + \square$
19. $6 = 4 + 1 + \square$
20. $6 = 2 + 4 + \square$

Unit 5: Relationship between Addition and Subtraction

Day 3B **Date:**

Why am I learning this?

Hailey wants $5 to buy a math workbook. If she earned $1 yesterday for doing chores, I can figure out how much more money Hailey needs to buy the math workbook.

1. $5 = 1 + \square$

2. $5 = 1 + 1 + \square$

3. $5 = 1 + 2 + \square$

4. $5 = 5 + \square$

5. $5 = 3 + \square$

6. $5 = 2 + 2 + \square$

7. $5 = 3 + 1 + \square$

8. $5 = 2 + \square$

9. $5 = 4 + \square$

10. $5 = 4 + 1 + \square$

11. $5 = 0 + 1 + \square$

12. $5 = 0 + \square$

13. $5 = 10 - \square$

14. $5 = 9 - 1 - \square$

15. $5 = 9 - 3 - \square$

16. $5 = 8 - \square$

17. $5 = 7 - \square$

18. $5 = 10 - 2 - \square$

19. $5 = 10 - 4 - \square$

20. $5 = 9 - \square$

Unit 5: Relationship between Addition and Subtraction

Day 4A **Date:**

Why am I learning this?

Levi's mom told Levi that he can only keep 4 tigers. If Levi has 10 tigers now, I can figure out how many more tigers that Levi needs to give away.

1. 4 = 10 - ☐
2. 4 = 10 - 6 - ☐
3. 4 = 10 - 1 - ☐
4. 4 = 9 - ☐
5. 4 = 8 - ☐
6. 4 = 10 - 5 - ☐
7. 4 = 10 - 2 - ☐
8. 4 = 7 - ☐
9. 4 = 6 - ☐
10. 4 = 10 - 4 - ☐
11. 4 = 10 - 5 - ☐
12. 4 = 5 - ☐
13. 4 = 4 - ☐
14. 4 = 9 - 3 - ☐
15. 4 = 9 - 4 - ☐
16. 4 = 9 - 1 - ☐
17. 4 = 7 - 1 - ☐
18. 4 = 8 - 3 - ☐
19. 4 = 8 - 1 - ☐
20. 4 = 10 - 3 - ☐

Unit 5: Relationship between Addition and Subtraction

Day 4B Date:

Why am I learning this?

Riley wants to save 3 hot dogs for her sister. If Riley has 10 hot dogs, I can figure out how many hot dogs that Riley can eat.

1. $3 = 10 - \square$
2. $3 = 10 - 1 - \square$
3. $3 = 10 - 5 - \square$
4. $3 = 9 - \square$
5. $3 = 8 - \square$
6. $3 = 10 - 4 - \square$
7. $3 = 10 - 2 - \square$
8. $3 = 7 - \square$
9. $3 = 6 - \square$
10. $3 = 10 - 6 - \square$
11. $3 = 9 - 5 - \square$
12. $3 = 5 - \square$
13. $3 = 4 - \square$
14. $3 = 9 - 2 - \square$
15. $3 = 9 - 3 - \square$
16. $3 = 3 - \square$
17. $3 = 9 - 6 - \square$
18. $3 = 8 - 3 - \square$
19. $3 = 8 - 1 - \square$
20. $3 = 7 - 2 - \square$

Unit 5: Relationship between Addition and Subtraction

Day 5A **Date:**

Why am I learning this?

Harison wants to keep only 2 castles. If Harison has 10 castles, I can figure out how many more castles Harison needs to give away.

1. 2 = 10 - ☐
2. 2 = 10 - 7 - ☐
3. 2 = 10 - 1 - ☐
4. 2 = 9 - ☐
5. 2 = 8 - ☐
6. 2 = 10 - 4 - ☐
7. 2 = 10 - 5 - ☐
8. 2 = 7 - ☐
9. 2 = 6 - ☐
10. 2 = 9 - 5 - ☐
11. 2 = 9 - 1 - ☐
12. 2 = 5 - ☐
13. 2 = 4 - ☐
14. 2 = 9 - 6 - ☐
15. 2 = 9 - 4 - ☐
16. 2 = 3 - ☐
17. 2 = 8 - 4 - ☐
18. 2 = 8 - 5 - ☐
19. 2 = 8 - 2 - ☐
20. 2 = 7 - 3 - ☐

Unit 5: Relationship between Addition and Subtraction

Day 5B Date:

Why am I learning this?

Amelia needs just 1 pencil for school. If Amelia has 10 pencils, I can figure out how many pencils Amelia can give away.

1. 1 = 10 - ☐

2. 1 = 10 - 1 - ☐

3. 1 = 10 - 5 - ☐

4. 1 = 9 - ☐

5. 1 = 8 - ☐

6. 1 = 10 - 7 - ☐

7. 1 = 10 - 3 - ☐

8. 1 = 7 - ☐

9. 1 = 6 - ☐

10. 1 = 9 - 7 - ☐

11. 1 = 9 - 2 - ☐

12. 1 = 5 - ☐

13. 1 = 4 - ☐

14. 1 = 9 - 4 - ☐

15. 1 = 9 - 6 - ☐

16. 1 = 3 - ☐

17. 1 = 8 - 5 - ☐

18. 1 = 8 - 3 - ☐

19. 1 = 7 - 2 - ☐

20. 1 = 6 - 4 - ☐

Finding a Missing Number

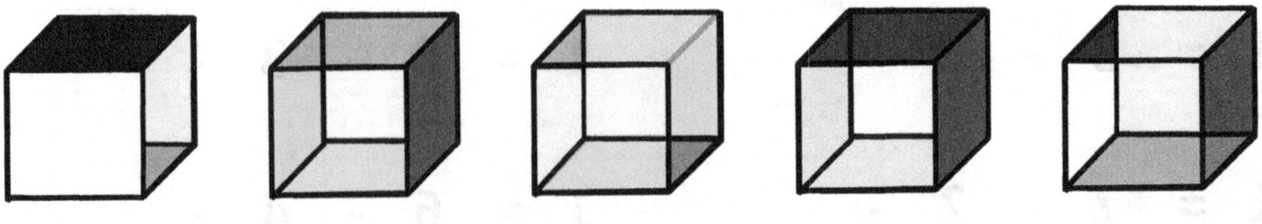

Unit 6: Finding a Missing Number

Lesson:

We use numbers every day to solve problems. In this unit, we will learn how to use numbers to solve problems like detectives!

Examples:

You had **8 watermelons**. You ate **4 watermelons** in the morning and ate **3 more watermelons** in the afternoon. How many watermelons are left? Can you figure it out?

8 – **4** – 3 = ?

❶ 8 – 4 = 4 ❷ 4 – 3 = 1

The answer is 1!

You have **2 apples**, **3 bananas**, and **4 pears**. How many fruits do you have in total? Can you figure it out?

2 + **3** + 4 = ?

❶ 2 + 3 = 5 ❷ 5 + 4 = 9

The answer is 9!

Unit 6: Finding a Missing Number

Day 1A **Date:**

Why am I learning this?

Sarah has 10 friends to invite to her birthday party, and she wrote 3 invitation cards. I can figure out how many more invitation cards that Sarah needs to write.

1. 10 - 3 = ☐
2. 8 - 3 - 3 = ☐
3. 9 - 5 = ☐
4. 10 - 2 - 4 = ☐
5. 9 - 3 = ☐
6. 6 - 1 - 1 = ☐
7. 8 - 2 = ☐
8. 9 - 3 - 1 = ☐
9. 4 - 2 = ☐
10. 6 - 2 - 2 = ☐
11. 10 - 7 = ☐
12. 9 - 3 - 1 = ☐
13. 9 - 4 = ☐
14. 6 - 1 - 1 = ☐
15. 10 - 4 = ☐
16. 7 - 1 - 3 = ☐
17. 8 - 5 = ☐
18. 6 - 3 - 3 = ☐
19. 5 - 2 = ☐
20. 8 - 5 - 2 = ☐

Unit 6: Finding a Missing Number

Day 1B Date:

Why am I learning this?

Cap'n Skully had 7 gold coins in a treasure chest earlier, but now he only has 4 gold coins in the treasure chest. I can figure out how many gold coins were stolen from the treasure chest.

1. 7 - 4 = ☐	2. 6 - 1 - 2 = ☐
3. 9 - 2 = ☐	4. 7 - 2 - 1 = ☐
5. 6 - 3 = ☐	6. 8 - 2 - 4 = ☐
7. 9 - 6 = ☐	8. 7 - 4 - 3 = ☐
9. 7 - 5 = ☐	10. 8 - 3 - 0 = ☐
11. 5 - 3 = ☐	12. 9 - 5 - 1 = ☐
13. 8 - 4 = ☐	14. 10 - 5 - 3 = ☐
15. 9 - 2 = ☐	16. 9 - 1 - 5 = ☐
17. 7 - 3 = ☐	18. 8 - 5 - 2 = ☐
19. 5 - 4 = ☐	20. 10 - 4 - 5 = ☐

Unit 6: Finding a Missing Number

Day 2A **Date:**

Why am I learning this?

Harper received $4 from her uncle, and $2 from her aunt for helping them. I can figure out how much money Harper received from her uncle and aunt.

1. 4 + 2 = ☐
2. 2 + 3 + 5 = ☐
3. 5 + 2 = ☐
4. 3 + 2 + 1 = ☐
5. 3 + 1 = ☐
6. 6 + 1 + 2 = ☐
7. 3 + 4 = ☐
8. 2 + 2 + 3 = ☐
9. 4 + 4 = ☐
10. 1 + 5 + 1 = ☐
11. 7 + 2 = ☐
12. 2 + 3 + 4 = ☐
13. 5 + 5 = ☐
14. 2 + 1 + 4 = ☐
15. 5 + 4 = ☐
16. 4 + 4 + 1 = ☐
17. 8 + 1 = ☐
18. 2 + 2 + 2 = ☐
19. 6 + 2 = ☐
20. 3 + 4 + 3 = ☐

Unit 6: Finding a Missing Number

Day 2B Date:

Why am I learning this?

Jordan had 8 rubies in his pocket, but lost 2 rubies because because his pocket had a hole. I can figure out how many rubies Jordan has left.

1. 8 - 2 = ☐
2. 5 - 2 - 3 = ☐
3. 9 - 4 = ☐
4. 6 - 2 - 2 = ☐
5. 7 - 2 = ☐
6. 8 - 0 - 2 = ☐
7. 8 - 3 = ☐
8. 10 - 4 - 1 = ☐
9. 10 - 3 = ☐
10. 9 - 3 - 3 = ☐
11. 7 - 4 = ☐
12. 9 - 5 - 2 = ☐
13. 9 - 7 = ☐
14. 7 - 2 - 5 = ☐
15. 5 - 3 = ☐
16. 8 - 2 - 6 = ☐
17. 7 - 6 = ☐
18. 10 - 2 - 4 = ☐
19. 6 - 4 = ☐
20. 9 - 1 - 7 = ☐

Unit 6: Finding a Missing Number

Day 3A **Date:**

Why am I learning this?

Jasmin would like to have 7 giraffes, and she has 5 giraffes already. I can figure out how many more giraffes Jasmin would like to get.

1. 7 - 5 = ☐	2. 4 - 1 - 2 = ☐
3. 9 - 3 = ☐	4. 10 - 3 - 4 = ☐
5. 8 - 2 = ☐	6. 9 - 2 - 3 = ☐
7. 5 - 3 = ☐	8. 9 - 2 - 6 = ☐
9. 9 - 5 = ☐	10. 8 - 1 - 4 = ☐
11. 10 - 6 = ☐	12. 8 - 3 - 2 = ☐
13. 9 - 6 = ☐	14. 7 - 2 - 2 = ☐
15. 9 - 4 = ☐	16. 8 - 2 - 5 = ☐
17. 10 - 4 = ☐	18. 8 - 3 - 1 = ☐
19. 7 - 2 = ☐	20. 6 - 2 - 3 = ☐

Unit 6: Finding a Missing Number

Day 3B Date:

Why am I learning this?

At a bus stop, 10 people waited for a bus. A bus came but it was quite full so 2 out of 10 people could not get into the bus. I can figure out how many people got on the bus.

1. $10 - 2 = \square$
2. $9 - 1 - 3 = \square$
3. $9 - 5 = \square$
4. $7 - 4 - 1 = \square$
5. $6 - 1 = \square$
6. $9 - 2 - 1 = \square$
7. $7 - 5 = \square$
8. $8 - 2 - 3 = \square$
9. $5 - 1 = \square$
10. $10 - 4 - 2 = \square$
11. $7 - 2 = \square$
12. $8 - 3 - 3 = \square$
13. $7 - 3 = \square$
14. $7 - 2 - 5 = \square$
15. $9 - 4 = \square$
16. $10 - 3 - 5 = \square$
17. $9 - 3 = \square$
18. $8 - 1 - 4 = \square$
19. $8 - 2 = \square$
20. $5 - 2 - 3 = \square$

Day 4A Date:

Why am I learning this?

Anna and Maya have super powers - Anna can eat 4 donuts per second, and Maya can eat 3 donuts per second. I can figure out how many donuts that Anna and Maya can eat in one second!

1. $4 + 3 = \square$
2. $3 + 2 + 1 = \square$
3. $3 + 3 = \square$
4. $5 + 2 + 2 = \square$
5. $2 + 4 = \square$
6. $6 + 1 + 1 = \square$
7. $4 + 4 = \square$
8. $6 + 0 + 2 = \square$
9. $3 + 5 = \square$
10. $2 + 4 + 3 = \square$
11. $2 + 2 = \square$
12. $3 + 3 + 4 = \square$
13. $4 + 5 = \square$
14. $1 + 3 + 3 = \square$
15. $6 + 3 = \square$
16. $5 + 2 + 1 = \square$
17. $2 + 3 = \square$
18. $2 + 4 + 4 = \square$
19. $7 + 2 = \square$
20. $4 + 1 + 2 = \square$

Unit 6: Finding a Missing Number

Day 4B Date:

Why am I learning this?

Connor had 8 rocket ships and he sent 3 rocket ships to Mars to explore. I can figure out how many rocket ships Connor has left to send to Jupiter.

1. 8 - 3 = ☐	2. 7 - 2 - 3 = ☐
3. 4 - 2 = ☐	4. 10 - 2 - 4 = ☐
5. 9 - 4 = ☐	6. 5 - 1 - 2 = ☐
7. 5 - 2 = ☐	8. 10 - 5 - 2 = ☐
9. 9 - 6 = ☐	10. 7 - 4 - 1 = ☐
11. 10 - 6 = ☐	12. 8 - 3 - 1 = ☐
13. 7 - 2 = ☐	14. 6 - 1 - 2 = ☐
15. 4 - 4 = ☐	16. 9 - 3 - 5 = ☐
17. 8 - 7 = ☐	18. 9 - 2 - 3 = ☐
19. 8 - 4 = ☐	20. 7 - 2 - 2 = ☐

Unit 6: Finding a Missing Number

Day 5A **Date:**

Why am I learning this?

Mia has 7 broccoli flavored cookies. Mia gives 4 of the cookies to her mom. I can figure out how many cookies Mia has left for herself. Yum!

1. 7 - 4 = ☐	2. 10 - 2 - 3 = ☐
3. 5 - 1 = ☐	4. 8 - 3 - 4 = ☐
5. 6 - 5 = ☐	6. 7 - 3 - 2 = ☐
7. 9 - 6 = ☐	8. 10 - 5 - 2 = ☐
9. 9 - 2 = ☐	10. 8 - 1 - 3 = ☐
11. 5 - 2 = ☐	12. 10 - 7 - 3 = ☐
13. 9 - 4 = ☐	14. 6 - 2 - 2 = ☐
15. 8 - 2 = ☐	16. 7 - 1 - 5 = ☐
17. 9 - 3 = ☐	18. 10 - 4 - 4 = ☐
19. 8 - 5 = ☐	20. 6 - 1 - 3 = ☐

Unit 6: Finding a Missing Number

Day 5B Date:

Why am I learning this?

Earlier, Lucas saw 10 aliens coming down from a space ship into his house. But now Lucas has 4 aliens in his house. I can figure out how many aliens left Lucas's house since then.

1. 10 - 4 = ☐
2. 9 - 1 - 6 = ☐
3. 5 - 2 = ☐
4. 9 - 2 - 5 = ☐
5. 5 - 3 = ☐
6. 7 - 4 - 2 = ☐
7. 8 - 3 = ☐
8. 10 - 4 - 6 = ☐
9. 9 - 7 = ☐
10. 8 - 4 - 3 = ☐
11. 6 - 2 = ☐
12. 7 - 2 - 5 = ☐
13. 7 - 3 = ☐
14. 9 - 2 - 4 = ☐
15. 7 - 2 = ☐
16. 6 - 2 - 1 = ☐
17. 6 - 4 = ☐
18. 6 - 3 - 2 = ☐
19. 10 - 2 = ☐
20. 8 - 2 - 3 = ☐

2-Digit Number + 1-Digit Number (Part 1)

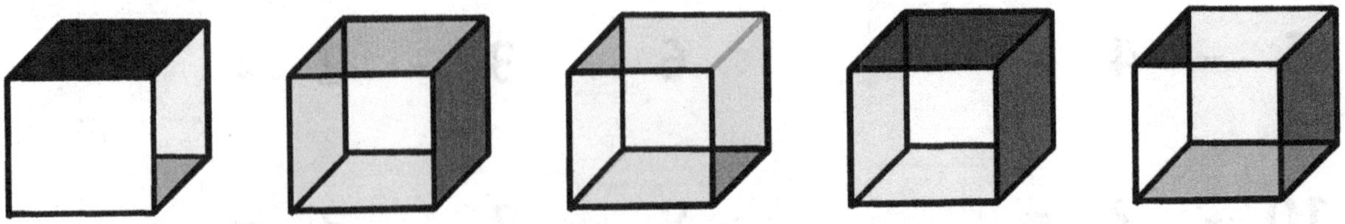

Unit 7: 2-Digit Number + 1-Digit Number (Part 1)

Lesson:

In this unit, we will learn to add a 2-digit number **(Tens)** and a 1-digit number **(Ones)**. For example, **24** is the same as **2 tens (20) and 4 ones (4)**.

Tens	Ones
2	4

= **24**

When you add a 2-digit number and a 1-digit number:

First, you need to **add the numbers in the ones place**, and write down the answer in the ones column on the bottom right.

Second, you need to **add the numbers in the tens place**, and write down the answer in the tens column on the bottom left. If there is no number in the tens place, you treat that as a zero.

Examples:

20 + 4 = ?

❶	Tens	Ones
	2	0
+		4
	?	?

❷	Tens	Ones
	2	0
+		4
		4

0 + 4 = 4

❸	Tens	Ones
	2	0
+		4
	2	4

2 + 0 = 2

OR

0 + 4 = 4

2 0 + 4 = 2 4

Day 1A Date:

Why am I learning this?

Emily has 30 freckles on her left cheek and 4 freckles on her right cheek. I can figure out how many freckles Emily has on her both cheeks.

1. 30 + 4 = ___
2. 8 + 20 = ___
3. 10 + 4 = ___
4. 4 + 50 = ___
5. 20 + 3 = ___
6. 3 + 90 = ___
7. 70 + 3 = ___
8. 2 + 50 = ___
9. 10 + 9 = ___
10. 7 + 70 = ___
11. 10 + 2 = ___
12. 5 + 30 = ___
13. 80 + 7 = ___
14. 3 + 80 = ___
15. 60 + 4 = ___
16. 0 + 90 = ___
17. 40 + 9 = ___
18. 6 + 20 = ___
19. 30 + 2 = ___
20. 9 + 10 = ___

Unit 7: 2-Digit Number + 1-Digit Number (Part 1)

Day 1B **Date:**

Why am I learning this?

Mason has 10 bubble gum pieces in his right pocket and 4 bubble gum pieces in his left pocket. I can figure out how many bubble gum pieces are in Mason's pockets.

1. 10 + 4 = ☐
2. 2 + 40 = ☐
3. 20 + 2 = ☐
4. 3 + 10 = ☐
5. 30 + 5 = ☐
6. 4 + 20 = ☐
7. 40 + 4 = ☐
8. 2 + 30 = ☐
9. 50 + 3 = ☐
10. 8 + 90 = ☐
11. 80 + 1 = ☐
12. 0 + 70 = ☐
13. 20 + 5 = ☐
14. 3 + 80 = ☐
15. 60 + 3 = ☐
16. 5 + 20 = ☐
17. 50 + 1 = ☐
18. 4 + 50 = ☐
19. 30 + 1 = ☐
20. 7 + 20 = ☐

Unit 7: 2-Digit Number + 1-Digit Number (Part 1)

Day 2A **Date:**

Why am I learning this?

While on vacation in Florida, Cindy spotted 20 mermaids on one day and saw 8 more mermaids on another day. I can figure out how many mermaids Cindy saw on vacation.

1.	2.	3.	4.	5.
20	4	60	8	90
+ 8	+ 50	+ 2	+ 20	+ 1

6.	7.	8.	9.	10.
3	50	1	50	6
+ 40	+ 9	+ 10	+ 8	+ 30

11.	12.	13.	14.	15.
20	9	60	7	50
+ 7	+ 20	+ 9	+ 70	+ 2

16.	17.	18.	19.	20.
8	50	2	10	3
+ 80	+ 7	+ 60	+ 5	+ 70

Unit 7: 2-Digit Number + 1-Digit Number (Part 1)

Day 2B **Date:**

Why am I learning this?

Wesley is wearing 30 layers of clothes because it is cold. Wesley put on 7 more layers because it is getting colder. I can figure out how layers of clothes that Wesley is wearing.

1. 30 + 7 = ☐
2. 9 + 50 = ☐
3. 10 + 9 = ☐
4. 4 + 40 = ☐
5. 50 + 2 = ☐
6. 9 + 80 = ☐
7. 70 + 0 = ☐
8. 2 + 10 = ☐
9. 80 + 4 = ☐
10. 5 + 60 = ☐
11. 30 + 3 = ☐
12. 3 + 40 = ☐
13. 90 + 6 = ☐
14. 0 + 10 = ☐
15. 30 + 6 = ☐
16. 6 + 40 = ☐
17. 20 + 8 = ☐
18. 1 + 70 = ☐
19. 40 + 3 = ☐
20. 5 + 20 = ☐

Unit 7: 2-Digit Number + 1-Digit Number (Part 1)

Day 3A **Date:**

Why am I learning this?

Jan needs to read two books. One book has 60 pages and the other book has 9 pages. I can figure out how many pages Jan has to read to finish both books.

1. 60 + 9 = ____
2. 2 + 80 = ____
3. 40 + 3 = ____
4. 5 + 60 = ____
5. 30 + 1 = ____
6. 6 + 20 = ____
7. 80 + 4 = ____
8. 2 + 60 = ____
9. 90 + 3 = ____
10. 5 + 50 = ____
11. 60 + 4 = ____
12. 3 + 20 = ____
13. 10 + 9 = ____
14. 6 + 70 = ____
15. 40 + 8 = ____
16. 7 + 30 = ____
17. 20 + 1 = ____
18. 8 + 90 = ____
19. 80 + 5 = ____
20. 3 + 50 = ____

Unit 7: 2-Digit Number + 1-Digit Number (Part 1)

Day 3B **Date:**

Why am I learning this?

Ted needs to spend 10 hours for math homework and 8 hours for science homework. I can figure out how many total hours Ted needs to spend on homework.

1. 10 + 8 = ☐
2. 0 + 80 = ☐
3. 80 + 3 = ☐
4. 3 + 30 = ☐
5. 30 + 1 = ☐
6. 7 + 10 = ☐
7. 50 + 4 = ☐
8. 8 + 50 = ☐
9. 70 + 6 = ☐
10. 9 + 90 = ☐
11. 20 + 7 = ☐
12. 4 + 20 = ☐
13. 40 + 9 = ☐
14. 6 + 40 = ☐
15. 90 + 2 = ☐
16. 2 + 70 = ☐
17. 50 + 3 = ☐
18. 7 + 90 = ☐
19. 80 + 7 = ☐
20. 1 + 60 = ☐

Day 4A Date:

Why am I learning this?

Michelle sells lemonade. On Saturday, Michelle sold 90 cups, and on Sunday, sold only 1 cup of lemonade. I can figure out how many cups of lemonade Michelle sold during the weekend.

1. $\begin{array}{r} 90 \\ +\ 1 \\ \hline \end{array}$

2. $\begin{array}{r} 4 \\ +\ 20 \\ \hline \end{array}$

3. $\begin{array}{r} 60 \\ +\ 4 \\ \hline \end{array}$

4. $\begin{array}{r} 2 \\ +\ 60 \\ \hline \end{array}$

5. $\begin{array}{r} 50 \\ +\ 8 \\ \hline \end{array}$

6. $\begin{array}{r} 5 \\ +\ 80 \\ \hline \end{array}$

7. $\begin{array}{r} 40 \\ +\ 2 \\ \hline \end{array}$

8. $\begin{array}{r} 7 \\ +\ 80 \\ \hline \end{array}$

9. $\begin{array}{r} 70 \\ +\ 2 \\ \hline \end{array}$

10. $\begin{array}{r} 9 \\ +\ 30 \\ \hline \end{array}$

11. $\begin{array}{r} 50 \\ +\ 2 \\ \hline \end{array}$

12. $\begin{array}{r} 1 \\ +\ 70 \\ \hline \end{array}$

13. $\begin{array}{r} 90 \\ +\ 6 \\ \hline \end{array}$

14. $\begin{array}{r} 4 \\ +\ 30 \\ \hline \end{array}$

15. $\begin{array}{r} 80 \\ +\ 9 \\ \hline \end{array}$

16. $\begin{array}{r} 7 \\ +\ 60 \\ \hline \end{array}$

17. $\begin{array}{r} 30 \\ +\ 4 \\ \hline \end{array}$

18. $\begin{array}{r} 5 \\ +\ 30 \\ \hline \end{array}$

19. $\begin{array}{r} 40 \\ +\ 5 \\ \hline \end{array}$

20. $\begin{array}{r} 7 \\ +\ 60 \\ \hline \end{array}$

Unit 7: 2-Digit Number + 1-Digit Number (Part 1)

Day 4B Date:

Why am I learning this?

Sean ate 60 chicken nuggets for breakfast and ate 8 chicken nuggets for lunch. I can figure out how many chicken nuggets total Sean ate for breakfast and lunch.

1. 60 + 8 = ☐	2. 5 + 20 = ☐
3. 70 + 4 = ☐	4. 7 + 30 = ☐
5. 30 + 2 = ☐	6. 7 + 40 = ☐
7. 80 + 6 = ☐	8. 3 + 10 = ☐
9. 90 + 0 = ☐	10. 6 + 50 = ☐
11. 10 + 3 = ☐	12. 1 + 70 = ☐
13. 50 + 2 = ☐	14. 9 + 0 = ☐
15. 80 + 7 = ☐	16. 7 + 50 = ☐
17. 40 + 5 = ☐	18. 4 + 30 = ☐
19. 90 + 2 = ☐	20. 2 + 80 = ☐

Unit 7: 2-Digit Number + 1-Digit Number (Part 1)

Day 5A **Date:**

Why am I learning this?

For her birthday, Sharon received 20 gifts from her friends. Sharon received 4 more gifts from her grandparents. I can figure out how many gifts Sharon received for her birthday.

1. $20 + 4 =$ ____
2. $1 + 90 =$ ____
3. $50 + 6 =$ ____
4. $3 + 70 =$ ____
5. $30 + 2 =$ ____
6. $7 + 40 =$ ____
7. $10 + 5 =$ ____
8. $4 + 30 =$ ____
9. $60 + 7 =$ ____
10. $8 + 40 =$ ____
11. $70 + 9 =$ ____
12. $0 + 60 =$ ____
13. $20 + 8 =$ ____
14. $1 + 30 =$ ____
15. $60 + 4 =$ ____
16. $2 + 70 =$ ____
17. $40 + 2 =$ ____
18. $5 + 80 =$ ____
19. $10 + 3 =$ ____
20. $2 + 50 =$ ____

Unit 7: 2-Digit Number + 1-Digit Number (Part 1)

Day 5B Date:

Why am I learning this?

Edward found 10 frogs in his pond, and then he noticed that another new frog jumped into the pond this morning. I can figure out how many frogs are in the pond.

1. 10 + 1 = ☐
2. 9 + 10 = ☐
3. 90 + 2 = ☐
4. 8 + 20 = ☐
5. 20 + 3 = ☐
6. 7 + 30 = ☐
7. 80 + 4 = ☐
8. 6 + 40 = ☐
9. 30 + 5 = ☐
10. 5 + 50 = ☐
11. 70 + 6 = ☐
12. 4 + 60 = ☐
13. 40 + 7 = ☐
14. 3 + 70 = ☐
15. 60 + 8 = ☐
16. 2 + 80 = ☐
17. 50 + 9 = ☐
18. 1 + 90 = ☐
19. 10 + 4 = ☐
20. 0 + 50 = ☐

UNIT 8

2-Digit Number + 1-Digit Number (Part 2)

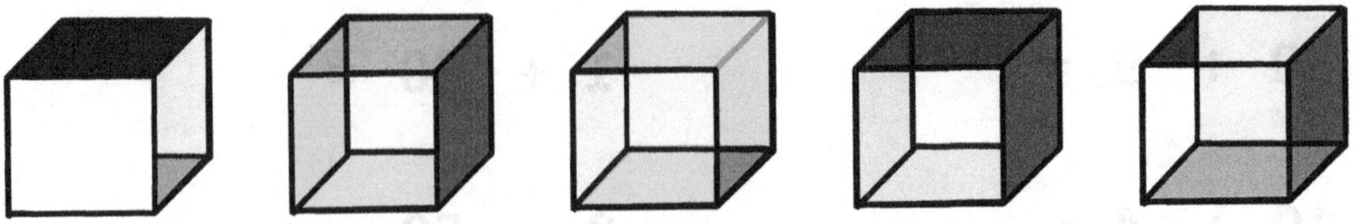

Unit 8: 2-Digit Number + 1-Digit Number (Part 2)

Lesson:

In this unit, we will learn how to add a 2-digit number that does not end with a zero and a 1-digit number.

When you add a 2-digit number and a 1-digit number:

First, you need to **add the numbers in the ones place**, and write down the answer in the ones column on the bottom right.

Second, you need to **add the numbers in the tens place**, and write down the answer in the tens column on the bottom left. If there is no number in the tens place, you treat that as a zero.

Examples:

7 + 31 = ?

❶	Tens	Ones
		7
+	3	1
	?	?

❷	Tens	Ones
		7
+	3	1
		8

7 + 1 = 8

❸	Tens	Ones
		7
+	3	1
	3	8

0 + 3 = 3

OR

7 + 1 = 8

7 + 3 1 = 3 8

Unit 8: 2-Digit Number + 1-Digit Number (Part 2)

Day 1A Date:

Why am I learning this?

Lyla had 25 holes in her pants this morning. During the recess, Lyla fell and made 2 more holes in her pants. I can figure out how many holes Lyla's pants have.

1. 25	2. 5	3. 21	4. 4	5. 52
+ 2	+ 52	+ 8	+ 12	+ 3
6. 3	7. 95	8. 2	9. 54	10. 5
+ 24	+ 3	+ 75	+ 2	+ 14
11. 72	12. 2	13. 33	14. 4	15. 84
+ 7	+ 12	+ 5	+ 82	+ 3
16. 6	17. 95	18. 2	19. 22	20. 2
+ 61	+ 2	+ 44	+ 6	+ 33

Day 1B Date:

Why am I learning this?

While hiking, George found 24 gold coins. Later, while riding a bike, George found 3 more gold coins. I can figure out how many gold coins George found without counting the coins individually.

1. 24 + 3 = ☐	2. 2 + 15 = ☐
3. 44 + 3 = ☐	4. 7 + 71 = ☐
5. 36 + 1 = ☐	6. 5 + 42 = ☐
7. 92 + 5 = ☐	8. 5 + 31 = ☐
9. 84 + 5 = ☐	10. 3 + 14 = ☐
11. 22 + 5 = ☐	12. 7 + 31 = ☐
13. 82 + 3 = ☐	14. 6 + 62 = ☐
15. 51 + 6 = ☐	16. 3 + 52 = ☐
17. 44 + 2 = ☐	18. 7 + 42 = ☐
19. 73 + 3 = ☐	20. 6 + 23 = ☐

Unit 8: 2-Digit Number + 1-Digit Number (Part 2)

Day 2A **Date:**

Why am I learning this?

73 cows are in Farmer John's pasture. 2 more cows walked into Farmer John's pasture. I can figure out how many cows are in Farmer John's pasture without counting them one by one.

1. 73 + 2
2. 4 + 51
3. 64 + 2
4. 8 + 21
5. 97 + 1
6. 3 + 44
7. 72 + 6
8. 1 + 14
9. 54 + 2
10. 6 + 32
11. 21 + 7
12. 4 + 34
13. 63 + 5
14. 6 + 71
15. 55 + 2
16. 4 + 84
17. 51 + 7
18. 2 + 63
19. 34 + 5
20. 3 + 25

Unit 8: 2-Digit Number + 1-Digit Number (Part 2)

Day 2B Date:

Why am I learning this?

My mom put 42 slices of pickles on my burger and then I added 3 more pickles. I can figure out how many slices of pickles are on my burger without counting them one by one.

1. 42 + 3 = ☐
2. 5 + 22 = ☐
3. 15 + 3 = ☐
4. 1 + 76 = ☐
5. 54 + 2 = ☐
6. 6 + 43 = ☐
7. 74 + 2 = ☐
8. 3 + 13 = ☐
9. 82 + 7 = ☐
10. 3 + 43 = ☐
11. 34 + 3 = ☐
12. 5 + 62 = ☐
13. 92 + 6 = ☐
14. 2 + 12 = ☐
15. 33 + 6 = ☐
16. 4 + 84 = ☐
17. 23 + 5 = ☐
18. 4 + 42 = ☐
19. 41 + 3 = ☐
20. 2 + 56 = ☐

Unit 8: 2-Digit Number + 1-Digit Number (Part 2)

Day 3A Date:

Why am I learning this?

Dylan walked for 42 miles in the morning and then walked an additional 7 miles that afternoon. I can figure out how many miles total Dylan walked that day.

1.

	4	2
+		7

2.

		2
+	9	3

3.

	4	6
+		3

4.

		2
+	6	3

5.

	3	7
+		1

6.

		3
+	2	4

7.

	8	5
+		4

8.

		2
+	6	2

9.

	9	7
+		2

10.

		5
+	5	3

11.

	6	4
+		4

12.

		3
+	2	3

13.

	1	2
+		4

14.

		6
+	7	1

15.

	4	2
+		6

16.

		2
+	3	3

17.

	2	4
+		1

18.

		6
+	9	2

19.

	8	3
+		5

20.

		3
+	5	6

Day 3B Date:

Why am I learning this?

We had strange weather today! Candy canes fell from the sky. Willow collected 46 candy canes, and Eli collected 3 candy canes. I can figure out how many candy canes Willow and Eli collected.

1. 46 + 3 = ☐
2. 7 + 91 = ☐
3. 84 + 3 = ☐
4. 2 + 72 = ☐
5. 33 + 1 = ☐
6. 6 + 43 = ☐
7. 52 + 4 = ☐
8. 1 + 65 = ☐
9. 72 + 6 = ☐
10. 4 + 23 = ☐
11. 22 + 7 = ☐
12. 4 + 93 = ☐
13. 45 + 2 = ☐
14. 2 + 56 = ☐
15. 93 + 3 = ☐
16. 1 + 15 = ☐
17. 53 + 2 = ☐
18. 3 + 35 = ☐
19. 82 + 7 = ☐
20. 4 + 82 = ☐

Unit 8: 2-Digit Number + 1-Digit Number (Part 2)

Day 4A Date:

Why am I learning this?

My dad planted 82 peach trees in our garden last week. My mom planted 7 more peach trees in our garden this week. I can figure out how many total peach trees my parents planted in our garden.

1.	2.	3.	4.	5.
82	8	14	4	25
+ 7	+ 21	+ 4	+ 52	+ 3

6.	7.	8.	9.	10.
3	74	2	15	7
+ 91	+ 3	+ 56	+ 2	+ 71

11.	12.	13.	14.	15.
17	5	81	3	65
+ 2	+ 33	+ 7	+ 84	+ 4

16.	17.	18.	19.	20.
4	47	4	32	6
+ 92	+ 1	+ 24	+ 2	+ 12

Unit 8: 2-Digit Number + 1-Digit Number (Part 2)

Day 4B Date:

Why am I learning this?

Roman ate 82 hotdogs for lunch today. Because he was so full, Roman ate only 5 hotdogs for dinner later that day. I can figure out how many hotdogs Roman ate that day.

1. 82 + 5 = ☐	2. 4 + 53 = ☐
3. 22 + 2 = ☐	4. 5 + 24 = ☐
5. 33 + 3 = ☐	6. 4 + 84 = ☐
7. 41 + 4 = ☐	8. 2 + 75 = ☐
9. 56 + 3 = ☐	10. 6 + 11 = ☐
11. 91 + 7 = ☐	12. 2 + 33 = ☐
13. 44 + 1 = ☐	14. 4 + 22 = ☐
15. 83 + 4 = ☐	16. 3 + 15 = ☐
17. 23 + 5 = ☐	18. 4 + 45 = ☐
19. 62 + 7 = ☐	20. 6 + 22 = ☐

Unit 8: 2-Digit Number + 1-Digit Number (Part 2)

Day 5A **Date:**

Why am I learning this?

Rose loves gardening. She grew 74 cantaloupes and 5 watermelons in her garden. I can figure out how many cantaloupes and watermelons Rose has total in her garden.

1. $74 + 5 =$ ___

2. $4 + 24 =$ ___

3. $52 + 6 =$ ___

4. $1 + 93 =$ ___

5. $36 + 2 =$ ___

6. $7 + 41 =$ ___

7. $13 + 5 =$ ___

8. $4 + 34 =$ ___

9. $62 + 6 =$ ___

10. $8 + 41 =$ ___

11. $73 + 3 =$ ___

12. $4 + 64 =$ ___

13. $25 + 2 =$ ___

14. $1 + 36 =$ ___

15. $52 + 4 =$ ___

16. $4 + 72 =$ ___

17. $43 + 2 =$ ___

18. $5 + 84 =$ ___

19. $14 + 3 =$ ___

20. $3 + 54 =$ ___

Unit 8: 2-Digit Number + 1-Digit Number (Part 2)

Day 5B

Date:

Why am I learning this?

Mila worked on math workbooks for 22 hours yesterday. Mila worked on math workbooks for 3 more hours today. I can figure out how long Mila practiced on workbooks yesterday and today.

1. 22 + 3 = ☐	2. 2 + 55 = ☐
3. 94 + 2 = ☐	4. 1 + 98 = ☐
5. 25 + 3 = ☐	6. 3 + 82 = ☐
7. 85 + 4 = ☐	8. 3 + 72 = ☐
9. 33 + 5 = ☐	10. 4 + 64 = ☐
11. 72 + 6 = ☐	12. 5 + 54 = ☐
13. 42 + 7 = ☐	14. 6 + 41 = ☐
15. 62 + 4 = ☐	16. 7 + 32 = ☐
17. 54 + 3 = ☐	18. 2 + 25 = ☐
19. 14 + 2 = ☐	20. 2 + 17 = ☐

2-Digit Number - 1-Digit Number

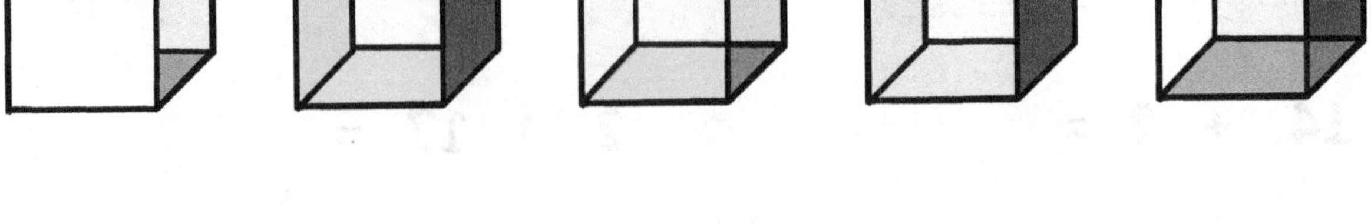

Unit 9: 2-Digit Number – 1-Digit Number (Part 1)

Lesson:

In this unit, we will learn how to subtract a 1-digit number from a 2-digit number.

When you subtract a 1-digit number from a 2-digit number:

First, you need to **subtract the numbers in the ones place**, and write down the answer in the ones column on the bottom right.

Second, you need to **subtract the numbers from the tens place**, and write down the answer in the tens column on the bottom left. If there is no number in the tens place, you treat that as a zero.

Examples:

49 – 7 = ?

❶	Tens	Ones
	4	9
–		7
	?	?

❷	Tens	Ones
	4	9
–		7
		2

9 – 7 = 2

❸	Tens	Ones
	4	9
–		7
	4	2

4 – 0 = 4

OR

9 – 7 = 2

4 9 – 7 = 4 2

Unit 9: 2-Digit Number – 1-Digit Number

Day 1A **Date:**

Why am I learning this?

Noah had 37 pet flies, but 3 flew away. I can figure out how many pet flies Noah has now without counting them one at a time.

1. $37 - 3$	2. $58 - 2$	3. $85 - 5$
4. $29 - 5$	5. $79 - 4$	6. $88 - 7$
7. $39 - 1$	8. $75 - 4$	9. $24 - 3$
10. $19 - 8$	11. $72 - 2$	12. $17 - 5$
13. $55 - 4$	14. $82 - 0$	15. $36 - 5$
16. $69 - 6$	17. $27 - 4$	18. $44 - 3$
19. $68 - 6$	20. $36 - 2$	

Unit 9: 2-Digit Number – 1-Digit Number

Day 1B Date:

Why am I learning this?

Quinn had a fish tank with 99 tropical fish. Quinn's cat ate 6 tropical fish before Quinn stopped the cat. I can figure out how many tropical fish are left in the tank.

1. 99 - 6 = ☐	2. 58 - 6 = ☐
3. 88 - 4 = ☐	4. 64 - 3 = ☐
5. 77 - 2 = ☐	6. 49 - 4 = ☐
7. 66 - 6 = ☐	8. 87 - 5 = ☐
9. 55 - 4 = ☐	10. 95 - 3 = ☐
11. 33 - 1 = ☐	12. 86 - 4 = ☐
13. 22 - 0 = ☐	14. 78 - 6 = ☐
15. 28 - 5 = ☐	16. 65 - 3 = ☐
17. 39 - 8 = ☐	18. 59 - 6 = ☐
19. 16 - 3 = ☐	20. 47 - 3 = ☐

Unit 9: 2-Digit Number – 1-Digit Number

Day 2A **Date:**

Why am I learning this?

Skylar baked 47 anchovy cookies. However, her younger brother ate 6 of her cookies while Skylar was not looking. I can figure out how many delicious anchovy cookies are left.

1. 47 − 6	2. 99 − 4	3. 67 − 5	4. 14 − 4	5. 29 − 5
6. 86 − 3	7. 74 − 2	8. 38 − 4	9. 88 − 7	10. 59 − 6
11. 94 − 4	12. 53 − 2	13. 17 − 5	14. 59 − 7	15. 25 − 3
16. 77 − 4	17. 28 − 5	18. 57 − 3	19. 65 − 2	20. 54 − 3

Unit 9: 2-Digit Number – 1-Digit Number

Day 2B **Date:**

Why am I learning this?

Jackson had 88 sports cars. One night, a dragon swooshed down and burned 4 of Jackson's cars. I can figure out how many sports cars Jackson has left.

1. 88 - 4 = ☐	2. 47 - 6 = ☐
3. 29 - 7 = ☐	4. 47 - 7 = ☐
5. 47 - 2 = ☐	6. 15 - 0 = ☐
7. 35 - 3 = ☐	8. 96 - 3 = ☐
9. 58 - 4 = ☐	10. 85 - 2 = ☐
11. 69 - 7 = ☐	12. 48 - 5 = ☐
13. 48 - 6 = ☐	14. 64 - 2 = ☐
15. 36 - 4 = ☐	16. 38 - 6 = ☐
17. 13 - 2 = ☐	18. 79 - 4 = ☐
19. 28 - 6 = ☐	20. 65 - 3 = ☐

Unit 9: 2-Digit Number – 1-Digit Number

Day 3A **Date:**

Why am I learning this?

Luna had 57 tiaras filled with pearls and diamonds. One night, a ninja came into Luna's castle and stole 7 of her tiaras. I can figure out how many tiaras are left for Luna.

1. 57 − 7 = ____
2. 27 − 4 = ____
3. 54 − 1 = ____
4. 36 − 4 = ____
5. 84 − 0 = ____
6. 66 − 4 = ____
7. 38 − 6 = ____
8. 29 − 6 = ____
9. 57 − 4 = ____
10. 76 − 3 = ____
11. 98 − 5 = ____
12. 65 − 2 = ____
13. 36 − 1 = ____
14. 48 − 2 = ____
15. 69 − 7 = ____
16. 36 − 5 = ____
17. 55 − 2 = ____
18. 27 − 2 = ____
19. 15 − 3 = ____
20. 28 − 4 = ____

Unit 9: 2-Digit Number – 1-Digit Number

Day 3B **Date:**

Why am I learning this?

A magic genie came and told Hunter that he can have 24 wishes. Hunter used 3 wishes by asking for pancakes, waffles, and bacon. I can figure out how many more wishes Hunter has left.

1. 24 - 3 = ☐
2. 36 - 2 = ☐
3. 19 - 3 = ☐
4. 25 - 1 = ☐
5. 84 - 2 = ☐
6. 34 - 0 = ☐
7. 57 - 4 = ☐
8. 58 - 6 = ☐
9. 38 - 7 = ☐
10. 69 - 8 = ☐
11. 77 - 2 = ☐
12. 97 - 5 = ☐
13. 95 - 4 = ☐
14. 85 - 3 = ☐
15. 64 - 2 = ☐
16. 46 - 5 = ☐
17. 27 - 5 = ☐
18. 98 - 6 = ☐
19. 59 - 6 = ☐
20. 74 - 2 = ☐

Unit 9: 2-Digit Number – 1-Digit Number

Day 4A **Date:**

Why am I learning this?

Sophia had 38 pairs of shoes. But her family adopted a puppy and the puppy chewed up 7 pairs of Sophia's shoes. I can figure out how many pairs of shoes Sophia has left.

1. 38 − 7 = ___
2. 96 − 5 = ___
3. 63 − 1 = ___
4. 27 − 4 = ___
5. 29 − 6 = ___
6. 48 − 5 = ___
7. 76 − 4 = ___
8. 62 − 2 = ___
9. 47 − 1 = ___
10. 25 − 3 = ___
11. 19 − 8 = ___
12. 35 − 3 = ___
13. 46 − 4 = ___
14. 87 − 1 = ___
15. 99 − 2 = ___
16. 75 − 2 = ___
17. 61 − 0 = ___
18. 14 − 1 = ___
19. 35 − 2 = ___
20. 59 − 5 = ___

Day 4B Date:

Why am I learning this?

Carter had 79 markers in his room. A blue fuzzy monster came into Carter's room and ate 4 of them. I can figure out how many markers Carter has left.

1. 79 - 4 = ☐

2. 64 - 3 = ☐

3. 98 - 6 = ☐

4. 55 - 1 = ☐

5. 32 - 1 = ☐

6. 86 - 2 = ☐

7. 48 - 5 = ☐

8. 25 - 3 = ☐

9. 62 - 0 = ☐

10. 48 - 6 = ☐

11. 57 - 6 = ☐

12. 69 - 7 = ☐

13. 54 - 2 = ☐

14. 14 - 2 = ☐

15. 17 - 4 = ☐

16. 36 - 5 = ☐

17. 28 - 5 = ☐

18. 47 - 1 = ☐

19. 37 - 7 = ☐

20. 65 - 2 = ☐

Unit 9: 2-Digit Number – 1-Digit Number

Day 5A **Date:**

Why am I learning this?

Rebecca is starting to read a 75-page book about a friendly ghost. Today Rebecca read 4 pages. I can figure out how many more pages Rebecca has left to finish the book.

1. $75 - 4 =$

2. $93 - 1 =$

3. $64 - 3 =$

4. $77 - 5 =$

5. $15 - 2 =$

6. $39 - 8 =$

7. $24 - 4 =$

8. $26 - 3 =$

9. $48 - 6 =$

10. $98 - 7 =$

11. $76 - 2 =$

12. $18 - 4 =$

13. $55 - 3 =$

14. $86 - 5 =$

15. $63 - 1 =$

16. $58 - 7 =$

17. $36 - 4 =$

18. $47 - 2 =$

19. $88 - 5 =$

20. $24 - 4 =$

Unit 9: 2-Digit Number – 1-Digit Number

Day 5B **Date:**

Why am I learning this?

Ada used to speak 24 languages. Because Ada did not practice, she forgot how to speak Spanish, French, and English. We can figure out how many languages Ada still knows how to speak.

1. 24 - 3 = ☐
2. 36 - 5 = ☐
3. 94 - 1 = ☐
4. 27 - 1 = ☐
5. 69 - 6 = ☐
6. 59 - 5 = ☐
7. 86 - 4 = ☐
8. 99 - 0 = ☐
9. 73 - 1 = ☐
10. 87 - 6 = ☐
11. 27 - 6 = ☐
12. 73 - 3 = ☐
13. 16 - 3 = ☐
14. 26 - 2 = ☐
15. 47 - 5 = ☐
16. 18 - 5 = ☐
17. 68 - 6 = ☐
18. 37 - 4 = ☐
19. 98 - 3 = ☐
20. 78 - 5 = ☐

Addition and Subtraction Review

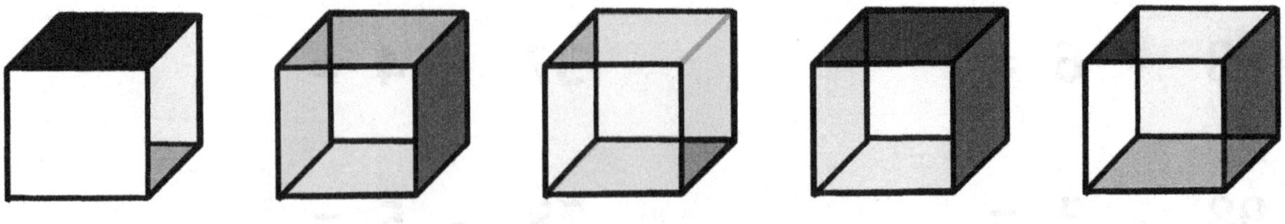

Unit 10: Addition and Subtraction Review

Lesson:

In this unit, we will review addition and subtraction. We are practicing to be accurate and quick in our calculations.

Did you know that practice is the only way to get better at something? Think about how you learned to walk, run, ride a bike, and read. Practicing is the only way!

Examples:

Addition:

6 + 53 = ?

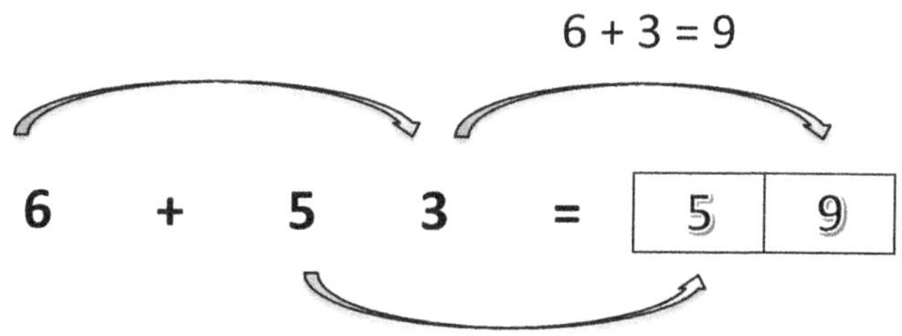

Subtraction:

29 – 5 = ?

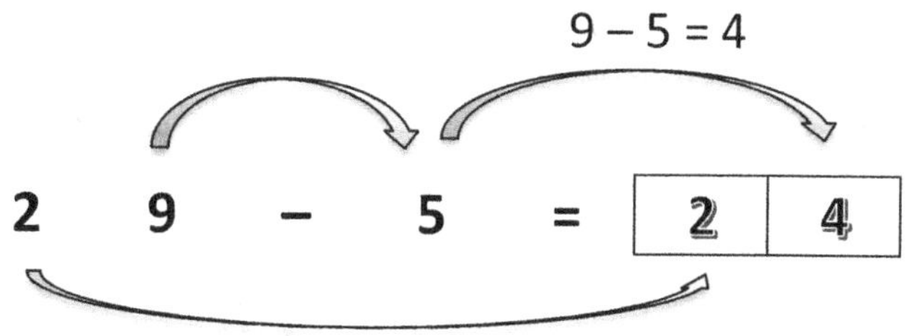

Unit 10: Addition and Subtraction Review

Day 1A **Date:**

Why am I learning this?

7 frogs were sitting on a log. 2 more frogs hopped on the log. I can figure out the total number of frogs on the log now.

1. 7 + 2 = ☐
2. 9 - 3 = ☐
3. 12 + 3 = ☐
4. 19 - 6 = ☐
5. 47 + 2 = ☐
6. 75 - 2 = ☐
7. 4 + 3 = ☐
8. 18 - 7 = ☐
9. 53 + 6 = ☐
10. 69 - 9 = ☐
11. 34 + 3 = ☐
12. 24 - 3 = ☐
13. 52 + 4 = ☐
14. 59 - 5 = ☐
15. 30 + 4 = ☐
16. 77 - 4 = ☐
17. 27 + 2 = ☐
18. 19 - 8 = ☐
19. 6 + 4 = ☐
20. 49 - 3 = ☐

Unit 10: Addition and Subtraction Review

Day 1B **Date:**

Why am I learning this?

Evelyn had 74 cats, but 3 of her cats ran away. I can figure out how many cats Evelyn has now.

1. **74 - 3 =** ☐	2. **62 + 2 =** ☐
3. **48 - 7 =** ☐	4. **23 + 4 =** ☐
5. **8 - 4 =** ☐	6. **16 + 3 =** ☐
7. **37 - 5 =** ☐	8. **30 + 7 =** ☐
9. **89 - 6 =** ☐	10. **22 + 5 =** ☐
11. **55 - 3 =** ☐	12. **81 + 5 =** ☐
13. **27 - 2 =** ☐	14. **24 + 5 =** ☐
15. **29 - 7 =** ☐	16. **21 + 6 =** ☐
17. **67 - 6 =** ☐	18. **42 + 4 =** ☐
19. **99 - 4 =** ☐	20. **91 + 7 =** ☐

Day 2A Date:

Why am I learning this?

Tyler did 56 backflips yesterday. Tyler did better backflips today by doing 3 more backflips than yesterday. I can figure out how many backflips Tyler did today.

1. 56 + 3 = ☐	2. 86 - 2 = ☐
3. 60 + 3 = ☐	4. 98 - 4 = ☐
5. 3 + 7 = ☐	6. 39 - 7 = ☐
7. 33 + 5 = ☐	8. 37 - 2 = ☐
9. 52 + 2 = ☐	10. 95 - 3 = ☐
11. 4 + 4 = ☐	12. 9 - 3 = ☐
13. 73 + 2 = ☐	14. 77 - 4 = ☐
15. 61 + 8 = ☐	16. 66 - 2 = ☐
17. 14 + 2 = ☐	18. 38 - 7 = ☐
19. 25 + 4 = ☐	20. 27 - 5 = ☐

Day 2B Date:

Why am I learning this?

Clumsy the Elf made 94 toys, but he broke 1 of them by accidentally stepping on it. I can figure out how many toys Clumsy the Elf has now.

1. 94 - 1 = ☐ 2. 81 + 4 = ☐

3. 23 - 0 = ☐ 4. 16 + 2 = ☐

5. 57 - 4 = ☐ 6. 60 + 2 = ☐

7. 38 - 6 = ☐ 8. 32 + 4 = ☐

9. 95 - 5 = ☐ 10. 50 + 3 = ☐

11. 18 - 3 = ☐ 12. 11 + 7 = ☐

13. 69 - 2 = ☐ 14. 62 + 7 = ☐

15. 7 - 2 = ☐ 16. 40 + 5 = ☐

17. 87 - 6 = ☐ 18. 81 + 2 = ☐

19. 46 - 3 = ☐ 20. 44 + 3 = ☐

Unit 10: Addition and Subtraction Review

Day 3A **Date:**

Why am I learning this?

Snail and Sloth gathered nuts together. Snail gathered 3 nuts and Sloth gathered 6 nuts. I can figure out how many nuts Snail and Sloth gathered together.

1. 3 + 6 = ☐
2. 8 - 5 = ☐
3. 55 + 2 = ☐
4. 97 - 4 = ☐
5. 12 + 5 = ☐
6. 15 - 4 = ☐
7. 30 + 3 = ☐
8. 63 - 1 = ☐
9. 73 + 0 = ☐
10. 77 - 5 = ☐
11. 24 + 4 = ☐
12. 28 - 4 = ☐
13. 81 + 5 = ☐
14. 86 - 2 = ☐
15. 52 + 4 = ☐
16. 88 - 6 = ☐
17. 47 + 2 = ☐
18. 46 - 5 = ☐
19. 32 + 6 = ☐
20. 39 - 6 = ☐

Unit 10: Addition and Subtraction Review

Day 3B **Date:**

Why am I learning this?

18 customers came to a toy shop to buy kites. Because the shop only had 7 kites, the toy shop sold only 1 kite per customer. I can figure out how many customers could not buy a kite.

1. 18 - 7 = ☐	2. 15 + 3 = ☐
3. 23 - 2 = ☐	4. 22 + 4 = ☐
5. 56 - 3 = ☐	6. 42 + 3 = ☐
7. 99 - 7 = ☐	8. 52 + 7 = ☐
9. 38 - 6 = ☐	10. 33 + 4 = ☐
11. 26 - 6 = ☐	12. 27 + 2 = ☐
13. 47 - 3 = ☐	14. 72 + 6 = ☐
15. 78 - 4 = ☐	16. 11 + 6 = ☐
17. 14 - 2 = ☐	18. 88 + 1 = ☐
19. 85 - 3 = ☐	20. 51 + 7 = ☐

Day 4A Date:

Why am I learning this?

Hannah has a magic jar. Hannah put 21 jelly beans in the jar and found 7 more jelly beans in a jar. I can figure out how many jelly beans Hannah has now.

1. 21 + 7 = ☐	2. 87 - 7 = ☐
3. 32 + 5 = ☐	4. 38 - 6 = ☐
5. 5 + 3 = ☐	6. 9 - 4 = ☐
7. 14 + 4 = ☐	8. 15 - 3 = ☐
9. 62 + 6 = ☐	10. 97 - 4 = ☐
11. 73 + 5 = ☐	12. 77 - 5 = ☐
13. 23 + 4 = ☐	14. 28 - 5 = ☐
15. 41 + 6 = ☐	16. 48 - 4 = ☐
17. 52 + 3 = ☐	18. 59 - 6 = ☐
19. 63 + 2 = ☐	20. 97 - 3 = ☐

Unit 10: Addition and Subtraction Review

Day 4B **Date:**

Why am I learning this?

Leah likes grubs. Leah dug into the ground and found 39 grubs. Leah's pet bird ate 4 of the grubs. I can figure out how many grubs Leah has now.

1. 39 - 4 = ☐
2. 31 + 8 = ☐
3. 16 - 3 = ☐
4. 12 + 7 = ☐
5. 47 - 6 = ☐
6. 42 + 3 = ☐
7. 87 - 3 = ☐
8. 82 + 6 = ☐
9. 32 - 1 = ☐
10. 73 + 4 = ☐
11. 76 - 4 = ☐
12. 77 + 2 = ☐
13. 58 - 6 = ☐
14. 82 + 5 = ☐
15. 95 - 2 = ☐
16. 47 + 2 = ☐
17. 28 - 3 = ☐
18. 23 + 6 = ☐
19. 16 - 2 = ☐
20. 13 + 5 = ☐

Day 5A Date:

Why am I learning this?

Ethan is getting ready for a snowball fight. Ethan had 76 snow balls and made 3 more snow balls. I can figure out how many snowballs Ethan has now.

1. 76 + 3 = ☐
2. 27 - 4 = ☐
3. 93 + 5 = ☐
4. 97 - 6 = ☐
5. 20 + 9 = ☐
6. 95 - 2 = ☐
7. 37 + 2 = ☐
8. 39 - 7 = ☐
9. 63 + 3 = ☐
10. 86 - 4 = ☐
11. 42 + 5 = ☐
12. 48 - 3 = ☐
13. 85 + 4 = ☐
14. 88 - 2 = ☐
15. 62 + 5 = ☐
16. 76 - 2 = ☐
17. 34 + 4 = ☐
18. 47 - 4 = ☐
19. 51 + 7 = ☐
20. 58 - 3 = ☐

Unit 10: Addition and Subtraction Review

Day 5B **Date:**

Why am I learning this?

Audrey had $18, but she spent $6 of her money on a dictionary. I can figure out how much money Audrey has now.

1. 18 - 6 = ☐	2. 12 + 5 = ☐
3. 26 - 4 = ☐	4. 21 + 8 = ☐
5. 49 - 7 = ☐	6. 43 + 3 = ☐
7. 29 - 4 = ☐	8. 22 + 6 = ☐
9. 57 - 6 = ☐	10. 55 + 3 = ☐
11. 38 - 2 = ☐	12. 34 + 5 = ☐
13. 86 - 3 = ☐	14. 41 + 4 = ☐
15. 64 - 4 = ☐	16. 82 + 6 = ☐
17. 27 - 6 = ☐	18. 32 + 4 = ☐
19. 67 - 4 = ☐	20. 62 + 6 = ☐

BOOK 1 ANSWER KEY

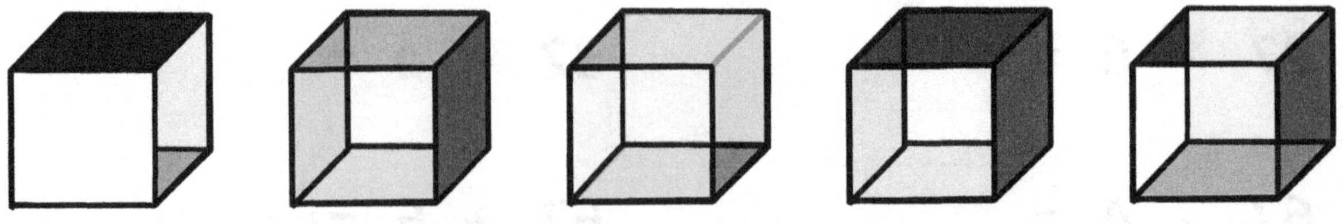

Answer Key - Unit 1

Unit 1, Day 1A (Page 8)

1.	**2**	2.	**4**	3.	**6**	4.	**8**	5.	**8**
6.	**4**	7.	**7**	8.	**7**	9.	**8**	10.	**6**
11.	**9**	12.	**6**	13.	**4**	14.	**6**	15.	**5**
16.	**8**	17.	**7**	18.	**9**	19.	**7**	20.	**8**

Unit 1, Day 1B (Page 9)

1.	**7**	2.	**4**	3.	**7**	4.	**8**	5.	**9**
6.	**7**	7.	**9**	8.	**8**	9.	**6**	10.	**8**
11.	**6**	12.	**8**	13.	**9**	14.	**6**	15.	**5**
16.	**8**	17.	**7**	18.	**9**	19.	**8**	20.	**7**

Unit 1, Day 2A (Page 10)

1.	**9**	2.	**9**	3.	**3**	4.	**9**	5.	**6**
6.	**6**	7.	**8**	8.	**8**	9.	**8**	10.	**8**
11.	**9**	12.	**3**	13.	**7**	14.	**8**	15.	**8**
16.	**8**	17.	**9**	18.	**8**	19.	**5**	20.	**5**

Unit 1, Day 2B (Page 11)

1.	**9**	2.	**6**	3.	**5**	4.	**6**	5.	**8**
6.	**8**	7.	**1**	8.	**9**	9.	**8**	10.	**7**
11.	**9**	12.	**8**	13.	**9**	14.	**9**	15.	**7**
16.	**9**	17.	**8**	18.	**7**	19.	**9**	20.	**6**

Unit 1, Day 3A (Page 12)

1.	**8**	2.	**8**	3.	**4**	4.	**9**	5.	**6**
6.	**2**	7.	**8**	8.	**4**	9.	**8**	10.	**5**
11.	**6**	12.	**6**	13.	**8**	14.	**7**	15.	**9**
16.	**7**	17.	**4**	18.	**6**	19.	**9**	20.	**8**

Unit 1, Day 3B (Page 13)

1.	**5**	2.	**7**	3.	**4**	4.	**9**	5.	**8**
6.	**7**	7.	**9**	8.	**9**	9.	**9**	10.	**6**
11.	**6**	12.	**6**	13.	**8**	14.	**9**	15.	**8**
16.	**8**	17.	**5**	18.	**8**	19.	**8**	20.	**5**

Unit 1, Day 4A (Page 14)

1.	**7**	2.	**8**	3.	**6**	4.	**9**	5.	**6**
6.	**7**	7.	**8**	8.	**6**	9.	**9**	10.	**6**
11.	**6**	12.	**9**	13.	**7**	14.	**5**	15.	**8**
16.	**6**	17.	**9**	18.	**7**	19.	**5**	20.	**8**

Unit 1, Day 4B (Page 15)

1.	**4**	2.	**7**	3.	**5**	4.	**9**	5.	**8**
6.	**7**	7.	**9**	8.	**9**	9.	**6**	10.	**8**
11.	**6**	12.	**8**	13.	**6**	14.	**4**	15.	**2**
16.	**9**	17.	**8**	18.	**6**	19.	**8**	20.	**9**

Unit 1, Day 5A (Page 16)

1.	**3**	2.	**9**	3.	**4**	4.	**5**	5.	**5**
6.	**8**	7.	**6**	8.	**6**	9.	**7**	10.	**7**
11.	**8**	12.	**7**	13.	**9**	14.	**6**	15.	**8**
16.	**4**	17.	**7**	18.	**5**	19.	**6**	20.	**9**

Unit 1, Day 5B (Page 17)

1.	**7**	2.	**9**	3.	**7**	4.	**9**	5.	**5**
6.	**9**	7.	**6**	8.	**8**	9.	**8**	10.	**4**
11.	**5**	12.	**6**	13.	**8**	14.	**6**	15.	**7**
16.	**5**	17.	**9**	18.	**8**	19.	**8**	20.	**9**

Answer Key - Unit 2

Unit 2, Day 1A (Page 20)									
1.	1	2.	6	3.	3	4.	0	5.	3
6.	4	7.	0	8.	1	9.	1	10.	5
11.	2	12.	2	13.	3	14.	4	15.	2
16.	1	17.	2	18.	0	19.	3	20.	1

Unit 2, Day 1B (Page 21)									
1.	2	2.	4	3.	5	4.	2	5.	7
6.	7	7.	5	8.	3	9.	1	10.	5
11.	4	12.	2	13.	1	14.	3	15.	4
16.	2	17.	1	18.	8	19.	3	20.	0

Unit 2, Day 2A (Page 22)									
1.	6	2.	2	3.	0	4.	2	5.	1
6.	3	7.	4	8.	2	9.	4	10.	3
11.	1	12.	5	13.	4	14.	5	15.	4
16.	4	17.	1	18.	2	19.	6	20.	2

Unit 2 Day 2B (Page 23)									
1.	5	2.	0	3.	2	4.	2	5.	3
6.	3	7.	1	8.	3	9.	4	10.	5
11.	6	12.	7	13.	3	14.	2	15.	3
16.	4	17.	1	18.	5	19.	4	20.	4

Unit 2, Day 3A (Page 24)									
1.	6	2.	4	3.	1	4.	3	5.	8
6.	1	7.	3	8.	0	9.	1	10.	4
11.	1	12.	4	13.	5	14.	2	15.	2
16.	2	17.	1	18.	6	19.	4	20.	1

Unit 2, Day 3B (Page 25)									
1.	4	2.	0	3.	1	4.	1	5.	3
6.	2	7.	0	8.	5	9.	3	10.	1
11.	6	12.	7	13.	3	14.	3	15.	3
16.	1	17.	2	18.	8	19.	3	20.	4

Unit 2, Day 4A (Page 26)									
1.	0	2.	2	3.	6	4.	1	5.	2
6.	3	7.	5	8.	4	9.	2	10.	0
11.	1	12.	7	13.	1	14.	7	15.	4
16.	4	17.	2	18.	2	19.	0	20.	5

Unit 2, Day 4B (Page 27)									
1.	3	2.	6	3.	3	4.	1	5.	3
6.	0	7.	3	8.	2	9.	1	10.	3
11.	1	12.	1	13.	0	14.	8	15.	2
16.	2	17.	2	18.	2	19.	6	20.	4

Unit 2, Day 5A (Page 28)									
1.	5	2.	5	3.	3	4.	4	5.	0
6.	1	7.	5	8.	2	9.	1	10.	3
11.	0	12.	3	13.	4	14.	4	15.	1
16.	2	17.	0	18.	1	19.	3	20.	3

Unit 2, Day 5B (Page 29)									
1.	5	2.	7	3.	1	4.	5	5.	0
6.	2	7.	5	8.	4	9.	2	10.	3
11.	7	12.	3	13.	2	14.	0	15.	2
16.	3	17.	0	18.	2	19.	4	20.	9

Answer Key - Unit 3

Unit 3, Day 1A (Page 32)			
1.	9	7	4
2.	7	3	3
3.	8	0	5
4.	6	4	5
5.	5	3	4
6.	9	1	5
7.	6	0	4
8.	9	1	7
9.	7	3	3
10.	8	1	2
11.	8	4	6
12.	6	1	2

Unit 3, Day 1B (Page 33)			
1.	4	9	7
2.	5	7	5
3.	1	8	6
4.	3	7	4
5.	3	5	4
6.	6	7	9
7.	1	8	5
8.	5	8	9
9.	6	7	5
10.	4	7	9
11.	3	5	8
12.	4	8	9

Unit 3, Day 2A (Page 34)			
1.	8	3	3
2.	9	1	2
3.	8	1	1
4.	7	4	2
5.	8	0	2
6.	3	1	3
7.	7	3	6
8.	7	2	5
9.	6	5	2
10.	8	2	3
11.	9	0	4
12.	5	4	2

Unit 3, Day 2B (Page 35)			
1.	5	8	8
2.	2	8	5
3.	3	4	9
4.	2	6	9
5.	4	2	6
6.	3	7	9
7.	1	8	3
8.	5	9	8
9.	2	4	7
10.	4	5	7
11.	6	5	9
12.	5	5	6

Unit 3, Day 3A (Page 36)			
1.	4	3	3
2.	8	6	5
3.	5	5	3
4.	7	7	5
5.	9	4	3
6.	9	4	0
7.	8	0	6
8.	6	1	4
9.	5	2	1
10.	6	2	1
11.	3	2	2
12.	5	3	1

Unit 3, Day 3B (Page 37)			
1.	4	7	6
2.	3	9	4
3.	3	8	4
4.	4	5	6
5.	1	5	3
6.	3	8	5
7.	5	7	9
8.	3	9	8
9.	2	7	8
10.	1	5	9
11.	5	9	6
12.	2	6	4

Unit 3, Day 4A (Page 38)			
1.	7	4	1
2.	9	2	1
3.	5	3	3
4.	4	0	5
5.	8	7	5
6.	9	1	4
7.	7	2	5
8.	8	0	4
9.	8	4	6
10.	9	4	2
11.	6	6	2
12.	8	1	7

Unit 3, Day 4B (Page 39)			
1.	7	9	8
2.	1	9	5
3.	5	8	7
4.	3	7	6
5.	1	4	8
6.	1	6	5
7.	6	8	9
8.	4	9	9
9.	2	7	4
10.	2	5	5
11.	1	3	6
12.	1	4	8

Unit 3, Day 5A (Page 40)			
1.	8	7	5
2.	9	0	4
3.	9	1	3
4.	5	2	1
5.	6	1	1
6.	8	2	6
7.	8	7	5
8.	9	2	8
9.	9	0	1
10.	9	6	5
11.	6	1	4
12.	7	4	3

Unit 3, Day 5B (Page 41)			
1.	3	5	5
2.	2	7	9
3.	2	2	9
4.	2	6	2
5.	3	5	8
6.	0	3	7
7.	1	4	7
8.	5	9	7
9.	6	8	8
10.	3	6	7
11.	1	8	5
12.	6	8	9

Answer Key - Unit 4

Unit 4, Day 1A (Page 44)

1.	6	2.	10
3.	9	4.	10
5.	8	6.	10
7.	9	8.	8
9.	8	10.	9
11.	10	12.	10
13.	10	14.	9
15.	9	16.	5
17.	8	18.	9
19.	9	20.	5

Unit 4, Day 1B (Page 45)

1.	4	2.	5
3.	0	4.	2
5.	2	6.	0
7.	2	8.	2
9.	2	10.	2
11.	2	12.	1
13.	1	14.	3
15.	1	16.	2
17.	3	18.	0
19.	4	20.	1

Unit 4, Day 2A (Page 46)

1.	9	2.	5
3.	9	4.	10
5.	4	6.	9
7.	3	8.	10
9.	5	10.	9
11.	9	12.	7
13.	7	14.	8
15.	8	16.	8
17.	7	18.	8
19.	9	20.	10

Unit 4, Day 2B (Page 47)

1.	4	2.	2
3.	3	4.	3
5.	0	6.	3
7.	1	8.	5
9.	3	10.	2
11.	2	12.	5
13.	5	14.	1
15.	1	16.	1
17.	6	18.	2
19.	3	20.	1

Unit 4, Day 3A (Page 48)

1.	9	2.	9
3.	6	4.	9
5.	8	6.	7
7.	10	8.	9
9.	7	10.	8
11.	4	12.	8
13.	8	14.	8
15.	7	16.	10
17.	9	18.	9
19.	7	20.	7

Unit 4, Day 3B (Page 49)

1.	4	2.	0
3.	1	4.	1
5.	4	6.	2
7.	1	8.	2
9.	2	10.	2
11.	1	12.	5
13.	2	14.	3
15.	3	16.	1
17.	1	18.	6
19.	1	20.	5

Unit 4, Day 4A (Page 50)

1.	3	2.	7
3.	9	4.	9
5.	9	6.	5
7.	8	8.	9
9.	10	10.	8
11.	9	12.	6
13.	5	14.	10
15.	6	16.	10
17.	4	18.	7
19.	7	20.	10

Unit 4, Day 4B (Page 51)

1.	2	2.	3
3.	2	4.	3
5.	1	6.	4
7.	4	8.	1
9.	0	10.	3
11.	2	12.	0
13.	1	14.	1
15.	2	16.	1
17.	3	18.	2
19.	1	20.	2

Unit 4, Day 5A (Page 52)

1.	7	2.	6
3.	10	4.	10
5.	7	6.	8
7.	8	8.	6
9.	7	10.	7
11.	10	12.	7
13.	6	14.	9
15.	7	16.	10
17.	8	18.	9
19.	9	20.	6

Unit 4, Day 5B (Page 53)

1.	2	2.	3
3.	2	4.	3
5.	1	6.	4
7.	4	8.	1
9.	0	10.	3
11.	1	12.	0
13.	1	14.	0
15.	2	16.	1
17.	2	18.	4
19.	1	20.	1

Answer Key - Unit 5

Unit 5, Day 1A (Page 56)			
1.	9	2.	3
3.	3	4.	7
5.	5	6.	2
7.	1	8.	3
9.	1	10.	2
11.	4	12.	8
13.	6	14.	3
15.	0	16.	4
17.	2	18.	1
19.	5	20.	10

Unit 5, Day 1B (Page 57)			
1.	8	2.	5
3.	0	4.	6
5.	4	6.	1
7.	3	8.	2
9.	7	10.	1
11.	2	12.	5
13.	3	14.	0
15.	3	16.	1
17.	9	18.	4
19.	2	20.	5

Unit 5, Day 2A (Page 58)			
1.	7	2.	1
3.	2	4.	4
5.	5	6.	1
7.	2	8.	2
9.	3	10.	6
11.	2	12.	8
13.	1	14.	5
15.	1	16.	7
17.	6	18.	3
19.	4	20.	0

Unit 5, Day 2B (Page 59)			
1.	6	2.	3
3.	4	4.	7
5.	4	6.	0
7.	3	8.	0
9.	2	10.	5
11.	1	12.	3
13.	5	14.	3
15.	1	16.	1
17.	1	18.	4
19.	0	20.	2

Unit 5, Day 3A (Page 60)			
1.	5	2.	4
3.	1	4.	6
5.	3	6.	3
7.	2	8.	0
9.	1	10.	2
11.	0	12.	0
13.	4	14.	2
15.	3	16.	4
17.	2	18.	1
19.	1	20.	0

Unit 5, Day 3B (Page 61)			
1.	4	2.	3
3.	2	4.	0
5.	2	6.	1
7.	1	8.	3
9.	1	10.	0
11.	4	12.	5
13.	5	14.	3
15.	1	16.	3
17.	2	18.	3
19.	1	20.	4

Unit 5, Day 4A (Page 62)			
1.	6	2.	0
3.	5	4.	5
5.	4	6.	1
7.	4	8.	3
9.	2	10.	2
11.	1	12.	1
13.	0	14.	2
15.	1	16.	4
17.	2	18.	1
19.	3	20.	3

Unit 5, Day 4B (Page 63)			
1.	7	2.	6
3.	2	4.	6
5.	5	6.	3
7.	5	8.	4
9.	3	10.	1
11.	1	12.	2
13.	1	14.	4
15.	3	16.	0
17.	0	18.	2
19.	4	20.	2

Unit 5, Day 5A (Page 64)			
1.	8	2.	1
3.	7	4.	7
5.	6	6.	4
7.	3	8.	5
9.	4	10.	2
11.	6	12.	3
13.	2	14.	1
15.	3	16.	1
17.	2	18.	1
19.	4	20.	2

Unit 5, Day 5B (Page 65)			
1.	9	2.	8
3.	4	4.	8
5.	7	6.	2
7.	6	8.	6
9.	5	10.	1
11.	6	12.	4
13.	3	14.	4
15.	2	16.	2
17.	2	18.	4
19.	4	20.	1

Answer Key - Unit 6

Unit 6, Day 1A (Page 68)

1.	7	2.	2
3.	4	4.	4
5.	6	6.	4
7.	6	8.	5
9.	2	10.	2
11.	3	12.	5
13.	5	14.	4
15.	6	16.	3
17.	3	18.	0
19.	3	20.	1

Unit 6, Day 1B (Page 69)

1.	3	2.	3
3.	7	4.	4
5.	3	6.	2
7.	3	8.	0
9.	2	10.	5
11.	2	12.	3
13.	4	14.	2
15.	7	16.	3
17.	4	18.	1
19.	1	20.	1

Unit 6, Day 2A (Page 70)

1.	6	2.	10
3.	7	4.	6
5.	4	6.	9
7.	7	8.	7
9.	8	10.	7
11.	9	12.	9
13.	10	14.	7
15.	9	16.	9
17.	9	18.	6
19.	8	20.	10

Unit 6, Day 2B (Page 71)

1.	6	2.	0
3.	5	4.	2
5.	5	6.	6
7.	5	8.	5
9.	7	10.	3
11.	3	12.	2
13.	2	14.	0
15.	2	16.	0
17.	1	18.	4
19.	2	20.	1

Unit 6, Day 3A (Page 72)

1.	2	2.	1
3.	6	4.	3
5.	6	6.	4
7.	2	8.	1
9.	4	10.	3
11.	4	12.	3
13.	3	14.	3
15.	5	16.	1
17.	6	18.	4
19.	5	20.	1

Unit 6, Day 3B (Page 73)

1.	8	2.	5
3.	4	4.	2
5.	5	6.	6
7.	2	8.	3
9.	4	10.	4
11.	5	12.	2
13.	4	14.	0
15.	5	16.	2
17.	6	18.	3
19.	6	20.	0

Unit 6, Day 4A (Page 74)

1.	7	2.	6
3.	6	4.	9
5.	6	6.	8
7.	8	8.	8
9.	8	10.	9
11.	4	12.	10
13.	9	14.	7
15.	9	16.	8
17.	5	18.	10
19.	9	20.	7

Unit 6, Day 4B (Page 75)

1.	5	2.	2
3.	2	4.	4
5.	5	6.	2
7.	3	8.	3
9.	3	10.	2
11.	4	12.	4
13.	5	14.	3
15.	0	16.	1
17.	1	18.	4
19.	4	20.	3

Unit 6, Day 5A (Page 76)

1.	3	2.	5
3.	4	4.	1
5.	1	6.	2
7.	3	8.	3
9.	7	10.	4
11.	3	12.	0
13.	5	14.	2
15.	6	16.	1
17.	6	18.	2
19.	3	20.	2

Unit 6, Day 5B (Page 77)

1.	6	2.	2
3.	3	4.	2
5.	2	6.	1
7.	5	8.	0
9.	2	10.	1
11.	4	12.	0
13.	4	14.	3
15.	5	16.	3
17.	2	18.	1
19.	8	20.	3

Answer Key - Unit 7

Unit 7, Day 1A (Page 80)									
1.	**34**	2.	**28**	3.	**14**	4.	**54**	5.	**23**
6.	**93**	7.	**73**	8.	**52**	9.	**19**	10.	**77**
11.	**12**	12.	**35**	13.	**87**	14.	**83**	15.	**64**
16.	**90**	17.	**49**	18.	**26**	19.	**32**	20.	**19**

Unit 7, Day 1B (Page 81)									
1.	**14**	2.	**42**	3.	**22**	4.	**13**	5.	**35**
6.	**24**	7.	**44**	8.	**32**	9.	**53**	10.	**98**
11.	**81**	12.	**70**	13.	**25**	14.	**83**	15.	**63**
16.	**25**	17.	**51**	18.	**54**	19.	**31**	20.	**27**

Unit 7, Day 2A (Page 82)									
1.	**28**	2.	**54**	3.	**62**	4.	**28**	5.	**91**
6.	**43**	7.	**59**	8.	**11**	9.	**58**	10.	**36**
11.	**27**	12.	**29**	13.	**69**	14.	**77**	15.	**52**
16.	**88**	17.	**57**	18.	**62**	19.	**15**	20.	**73**

Unit 7, Day 2B (Page 83)									
1.	**37**	2.	**59**	3.	**19**	4.	**44**	5.	**52**
6.	**89**	7.	**70**	8.	**12**	9.	**84**	10.	**65**
11.	**33**	12.	**43**	13.	**96**	14.	**10**	15.	**36**
16.	**46**	17.	**28**	18.	**71**	19.	**43**	20.	**25**

Unit 7, Day 3A (Page 84)									
1.	**69**	2.	**82**	3.	**43**	4.	**65**	5.	**31**
6.	**26**	7.	**84**	8.	**62**	9.	**93**	10.	**55**
11.	**64**	12.	**23**	13.	**19**	14.	**76**	15.	**48**
16.	**37**	17.	**21**	18.	**98**	19.	**85**	20.	**53**

Unit 7, Day 3B (Page 85)									
1.	**18**	2.	**80**	3.	**83**	4.	**33**	5.	**31**
6.	**17**	7.	**54**	8.	**58**	9.	**76**	10.	**99**
11.	**27**	12.	**24**	13.	**49**	14.	**46**	15.	**92**
16.	**72**	17.	**53**	18.	**97**	19.	**87**	20.	**61**

Unit 7, Day 4A (Page 86)									
1.	**91**	2.	**24**	3.	**64**	4.	**62**	5.	**58**
6.	**85**	7.	**42**	8.	**87**	9.	**72**	10.	**39**
11.	**52**	12.	**71**	13.	**96**	14.	**34**	15.	**89**
16.	**67**	17.	**34**	18.	**35**	19.	**45**	20.	**67**

Unit 7, Day 4B (Page 87)									
1.	**68**	2.	**25**	3.	**74**	4.	**37**	5.	**32**
6.	**47**	7.	**86**	8.	**13**	9.	**90**	10.	**56**
11.	**13**	12.	**71**	13.	**52**	14.	**9**	15.	**87**
16.	**57**	17.	**45**	18.	**34**	19.	**92**	20.	**82**

Unit 7, Day 5A (Page 88)									
1.	**24**	2.	**91**	3.	**56**	4.	**73**	5.	**32**
6.	**47**	7.	**15**	8.	**34**	9.	**67**	10.	**48**
11.	**79**	12.	**60**	13.	**28**	14.	**31**	15.	**64**
16.	**72**	17.	**42**	18.	**85**	19.	**13**	20.	**52**

Unit 7, Day 5B (Page 89)									
1.	**11**	2.	**19**	3.	**92**	4.	**28**	5.	**23**
6.	**37**	7.	**84**	8.	**46**	9.	**35**	10.	**55**
11.	**76**	12.	**64**	13.	**47**	14.	**73**	15.	**68**
16.	**82**	17.	**59**	18.	**91**	19.	**14**	20.	**50**

Answer Key - Unit 8

Unit 8, Day 1A (Page 92)

1.	27	2.	57	3.	29	4.	16	5.	55
6.	27	7.	98	8.	77	9.	56	10.	19
11.	79	12.	14	13.	38	14.	86	15.	87
16.	67	17.	97	18.	46	19.	28	20.	35

Unit 8, Day 1B (Page 93)

1.	27	2.	17	3.	47	4.	78	5.	37
6.	47	7.	97	8.	36	9.	89	10.	17
11.	27	12.	38	13.	85	14.	68	15.	57
16.	55	17.	46	18.	49	19.	76	20.	29

Unit 8, Day 2A (Page 94)

1.	75	2.	55	3.	66	4.	29	5.	98
6.	47	7.	78	8.	15	9.	56	10.	38
11.	28	12.	38	13.	68	14.	77	15.	57
16.	88	17.	58	18.	65	19.	39	20.	28

Unit 8, Day 2B (Page 95)

1.	45	2.	27	3.	18	4.	77	5.	56
6.	49	7.	76	8.	16	9.	89	10.	46
11.	37	12.	67	13.	98	14.	14	15.	39
16.	88	17.	28	18.	46	19.	44	20.	58

Unit 8, Day 3A (Page 96)

1.	49	2.	95	3.	49	4.	65	5.	38
6.	27	7.	89	8.	64	9.	99	10.	58
11.	68	12.	26	13.	16	14.	77	15.	48
16.	35	17.	25	18.	98	19.	88	20.	59

Unit 8, Day 3B (Page 97)

1.	49	2.	98	3.	87	4.	74	5.	34
6.	49	7.	56	8.	66	9.	78	10.	27
11.	29	12.	97	13.	47	14.	58	15.	96
16.	16	17.	55	18.	38	19.	89	20.	86

Unit 8, Day 4A (Page 98)

1.	89	2.	29	3.	18	4.	56	5.	28
6.	94	7.	77	8.	58	9.	17	10.	78
11.	19	12.	38	13.	88	14.	87	15.	69
16.	96	17.	48	18.	28	19.	34	20.	18

Unit 8, Day 4B (Page 99)

1.	87	2.	57	3.	24	4.	29	5.	36
6.	88	7.	45	8.	77	9.	59	10.	17
11.	98	12.	35	13.	45	14.	26	15.	87
16.	18	17.	28	18.	49	19.	69	20.	28

Unit 8, Day 5A (Page 100)

1.	79	2.	28	3.	58	4.	94	5.	38
6.	48	7.	18	8.	38	9.	68	10.	49
11.	76	12.	68	13.	27	14.	37	15.	56
16.	76	17.	45	18.	89	19.	17	20.	57

Unit 8, Day 5B (Page 101)

1.	25	2.	57	3.	96	4.	99	5.	28
6.	85	7.	89	8.	75	9.	38	10.	68
11.	78	12.	59	13.	49	14.	47	15.	66
16.	39	17.	57	18.	27	19.	16	20.	19

Answer Key - Unit 9

Unit 9, Day 1A (Page 104)

1.	**34**	2.	**56**	3.	**80**	4.	**24**	5.	**75**
6.	**81**	7.	**38**	8.	**71**	9.	**21**	10.	**11**
11.	**70**	12.	**12**	13.	**51**	14.	**82**	15.	**31**
16.	**63**	17.	**23**	18.	**41**	19.	**62**	20.	**34**

Unit 9, Day 1B (Page 105)

1.	**93**	2.	**52**	3.	**84**	4.	**61**	5.	**75**
6.	**45**	7.	**60**	8.	**82**	9.	**51**	10.	**92**
11.	**32**	12.	**82**	13.	**22**	14.	**72**	15.	**23**
16.	**62**	17.	**31**	18.	**53**	19.	**13**	20.	**44**

Unit 9, Day 2A (Page 106)

1.	**41**	2.	**95**	3.	**62**	4.	**10**	5.	**24**
6.	**83**	7.	**72**	8.	**34**	9.	**81**	10.	**53**
11.	**90**	12.	**51**	13.	**12**	14.	**52**	15.	**22**
16.	**73**	17.	**23**	18.	**54**	19.	**63**	20.	**51**

Unit 9, Day 2B (Page 107)

1.	**84**	2.	**41**	3.	**22**	4.	**40**	5.	**45**
6.	**15**	7.	**32**	8.	**93**	9.	**54**	10.	**83**
11.	**62**	12.	**43**	13.	**42**	14.	**62**	15.	**32**
16.	**32**	17.	**11**	18.	**75**	19.	**22**	20.	**62**

Unit 9, Day 3A (Page 108)

1.	**50**	2.	**23**	3.	**53**	4.	**32**	5.	**84**
6.	**62**	7.	**32**	8.	**23**	9.	**53**	10.	**73**
11.	**93**	12.	**63**	13.	**35**	14.	**46**	15.	**62**
16.	**31**	17.	**53**	18.	**25**	19.	**12**	20.	**24**

Unit 9, Day 3B (Page 109)

1.	**21**	2.	**34**	3.	**16**	4.	**24**	5.	**82**
6.	**34**	7.	**53**	8.	**52**	9.	**31**	10.	**61**
11.	**71**	12.	**92**	13.	**91**	14.	**82**	15.	**62**
16.	**41**	17.	**22**	18.	**92**	19.	**53**	20.	**72**

Unit 9, Day 4A (Page 110)

1.	**31**	2.	**91**	3.	**62**	4.	**23**	5.	**23**
6.	**43**	7.	**72**	8.	**60**	9.	**46**	10.	**22**
11.	**11**	12.	**32**	13.	**42**	14.	**86**	15.	**97**
16.	**73**	17.	**61**	18.	**13**	19.	**33**	20.	**54**

Unit 9, Day 4B (Page 111)

1.	**75**	2.	**61**	3.	**92**	4.	**54**	5.	**31**
6.	**84**	7.	**43**	8.	**22**	9.	**62**	10.	**42**
11.	**51**	12.	**62**	13.	**52**	14.	**12**	15.	**13**
16.	**31**	17.	**23**	18.	**46**	19.	**30**	20.	**63**

Unit 9, Day 5A (Page 112)

1.	**71**	2.	**92**	3.	**61**	4.	**72**	5.	**13**
6.	**31**	7.	**20**	8.	**23**	9.	**42**	10.	**91**
11.	**74**	12.	**14**	13.	**52**	14.	**81**	15.	**62**
16.	**51**	17.	**32**	18.	**45**	19.	**83**	20.	**20**

Unit 9, Day 5B (Page 113)

1.	**21**	2.	**31**	3.	**93**	4.	**26**	5.	**63**
6.	**54**	7.	**82**	8.	**99**	9.	**72**	10.	**81**
11.	**21**	12.	**70**	13.	**13**	14.	**24**	15.	**42**
16.	**13**	17.	**62**	18.	**33**	19.	**95**	20.	**73**

Answer Key - Unit 10

Unit 10, Day 1A (Page 116)			
1.	9	2.	6
3.	15	4.	13
5.	49	6.	73
7.	7	8.	11
9.	59	10.	60
11.	37	12.	21
13.	56	14.	54
15.	34	16.	73
17.	29	18.	11
19.	10	20.	46

Unit 10, Day 1B (Page 117)			
1.	71	2.	64
3.	41	4.	27
5.	4	6.	19
7.	32	8.	37
9.	83	10.	27
11.	52	12.	86
13.	25	14.	29
15.	22	16.	27
17.	61	18.	46
19.	95	20.	98

Unit 10, Day 2A (Page 118)			
1.	59	2.	84
3.	63	4.	94
5.	10	6.	32
7.	38	8.	35
9.	54	10.	92
11.	8	12.	6
13.	75	14.	73
15.	69	16.	64
17.	16	18.	31
19.	29	20.	22

Unit 10, Day 2B (Page 119)			
1.	93	2.	85
3.	23	4.	18
5.	53	6.	62
7.	32	8.	36
9.	90	10.	53
11.	15	12.	18
13.	67	14.	69
15.	5	16.	45
17.	81	18.	83
19.	43	20.	47

Unit 10, Day 3A (Page 120)			
1.	9	2.	3
3.	57	4.	93
5.	17	6.	11
7.	33	8.	62
9.	73	10.	72
11.	28	12.	24
13.	86	14.	84
15.	56	16.	82
17.	49	18.	41
19.	38	20.	33

Unit 10, Day 3B (Page 121)			
1.	11	2.	18
3.	21	4.	26
5.	53	6.	45
7.	92	8.	59
9.	32	10.	37
11.	20	12.	29
13.	44	14.	78
15.	74	16.	17
17.	12	18.	89
19.	82	20.	58

Unit 10, Day 4A (Page 122)			
1.	28	2.	80
3.	37	4.	32
5.	8	6.	5
7.	18	8.	12
9.	68	10.	93
11.	78	12.	72
13.	27	14.	23
15.	47	16.	44
17.	55	18.	53
19.	65	20.	94

Unit 10, Day 4B (Page 123)			
1.	35	2.	39
3.	13	4.	19
5.	41	6.	45
7.	84	8.	88
9.	31	10.	77
11.	72	12.	79
13.	52	14.	87
15.	93	16.	49
17.	25	18.	29
19.	14	20.	18

Unit 10, Day 5A (Page 124)			
1.	79	2.	23
3.	98	4.	91
5.	29	6.	93
7.	39	8.	32
9.	66	10.	82
11.	47	12.	45
13.	89	14.	86
15.	67	16.	74
17.	38	18.	43
19.	58	20.	55

Unit 10, Day 5B (Page 125)			
1.	12	2.	17
3.	22	4.	29
5.	42	6.	46
7.	25	8.	28
9.	51	10.	58
11.	36	12.	39
13.	83	14.	45
15.	60	16.	88
17.	21	18.	36
19.	63	20.	68

Made in the USA
Monee, IL
08 November 2026

67972903R00077